DECARBONISING ELECTRICITY MADE SIMPLE

This book assesses how low-carbon generation, the advance of energy storage and consumer-based models can help decarbonise electricity supplies at a national level.

This book is built around developing a decarbonised electricity mix for Britain which reduces fossil fuels from 50% of supply in 2018 down to levels within 2030 interim carbon goals on the way to net zero emissions. Crossland explores the idea of a future energy storage mix which blends domestic batteries, vehicles, thermal stores and pumped hydro to provide a flexible, responsive electricity system. He then goes on to look at how much storage can contribute to decarbonisation in a multitude of contexts – from domestic to national electricity. This book also discusses how efficiency and self-sufficiency can bring about a decarbonised electricity use within our homes today. Britain is used as the main example, but the themes and conclusions are applicable to a global audience, and each chapter draws on practical case studies from around the world to illustrate key ideas.

Drawing on the author's experience in delivering and analysing low-carbon energy projects in the UK, Sub-Saharan Africa, Latin America and Oceania, this book will be of great relevance to students, scholars and industry specialists with an interest in energy technology, policy and storage.

Andrew F. Crossland is a specialist in the social, technical and economic modelling of energy systems. In 2017, he won the "Rising Star" Award from Energy UK recognising some of his work in the rapidly changing electricity system including work on the platform MyGridGB. His PhD from Durham University focused on the techno-economic assessment of energy storage in integrating solar photovoltaic (PV) on electricity networks.

Routledge Explorations in Energy Studies

DECARBONISING ELECTRICITY MADE SIMPLE

Andrew F. Crossland

LONDON AND NEW YORK

from Routledge

First published 2020
by Routledge
2 Park Square, Milton Park, Abingdon, Oxon OX14 4RN

and by Routledge
52 Vanderbilt Avenue, New York, NY 10017

Routledge is an imprint of the Taylor & Francis Group, an informa business

British Library Cataloguing-in-Publication Data
A catalogue record for this book is available from the British Library

Library of Congress Cataloging-in-Publication Data
Names: Crossland, Andrew F., author.
Title: Decarbonising electricity made simple / Andrew F. Crossland.
Description: Abingdon, Oxon; New York, NY: Routledge, 2020. |
Series: Routledge explorations in energy studies |
Includes bibliographical references and index.
Identifiers: LCCN 2019018258 (print) | LCCN 2019018554 (ebook) |
ISBN 9780367203337 (Master) | ISBN 9780367203313
(hardback: alk. paper) | ISBN 9780367203320 (pbk.: alk. paper) |
ISBN 9780367203337 (ebook)
Subjects: LCSH: Electric power production—Technological innovations. |
Renewable energy sources. | Carbon dioxide mitigation. |
Electric power-plants—Fuel.
Classification: LCC TK1005 (ebook) | LCC TK1005 .C755 2020
(print) | DDC 333.79/4—dc23
LC record available at https://lccn.loc.gov/2019018258

ISBN: 978-0-367-20331-3 (hbk)
ISBN: 978-0-367-20332-0 (pbk)
ISBN: 978-0-367-20333-7 (ebk)

Typeset in Bembo
by codeMantra

For Mum and Dad

CONTENTS

FIGURES

TABLES

ACKNOWLEDGEMENTS

It's inspiring working with organisations who are presently working hard on delivering low-carbon energy systems around the world. I am permanently indebted to the amazing people who have sparked low-carbon thinking at Durham University, Newcastle University, Durham Energy Institute, Loughborough University, Electricity North West, SolarCentury, Transpower, Network Rail, BRE, ACE, Raleigh International, Advance Further Energy and my present employer, Infratec (who really are a game-changing company). That gratitude extends from the engineers, lawyers, sales teams, developers, PR teams, admin and academics who've given me your time and shown me your patience—thank you.

Huge gratitude is extended to the various reviewers and supporters of this book and myself through my career, particularly my amazing adopted sister Rukiya "Rafiki" Sood and various colleagues, family and friends including Jan Muller, Dr Allen Wang, Dr Harold Anuta, Eman, Dr Neal Wade, David Minnis, Keith Scoles, Michael Richardson, Fiona Baddeley, Barbara Denz, Margaret Walker, Dr Grant Wilson, Cliff Willson, Charles Fletcher and Dr Philip Leicester. Thank you all.

Above all, throughout my life, there are two people who have guided and supported me beyond anyone else. Mum and Dad, you are my rock, my strength and my inspiration. I hope that all I do will make you as proud of me as I am of you.

PREFACE

In November 2011, I attended a talk by the late Professor David MacKay at a conference in London. Less than three years after releasing his book "Sustainable Energy without the Hot Air," Professor Mackay was helping to develop the UK Government's low-carbon strategy and talking about a new energy calculator he had built. At the time, the UK was beginning to enact policies to try to reduce British dependence on fossil fuels and to lower greenhouse gas emissions. Interventions like the Feed-in-Tariff and Renewables Obligation Certificates were providing generous and stable returns for investors in wind, solar, biomass and hydroelectric power plants, yet renewable technologies were barely contributing to the energy mix.

Sitting in the audience at the conference, I was impressed by Professor MacKay. His analytical approach and application of his science towards the issue of sustainability was inspiring. I remember how during his presentation he confronted both the strengths and flaws in his book and argued that it was better to put evidence in front of people for informed debate rather than risking not debating at all. From that moment, I have been inspired by his work on forwarding understanding of energy and decarbonisation. He saw the value in putting real numbers into a debate which was often misinformed by a biased or misleading media and some bad science. My PhD would never have the impact that the late Professor MacKay had, but his talk that day has always resonated with me.

Moving forward to 2015, I finished my PhD in renewable electricity and energy storage and was starting a career as an electrical engineer. When people realised that I spent three years researching sustainable energy, they would usually ask me what I thought about the low-carbon transition. It struck me that although there was wide interest in sustainability at the time, people were not empowered with the information they needed to form rational decisions on our energy future. Misinformation was rife, and it seemed unfair and dangerous that the public were being misled by all sides of such an important and popular topic.

I thought that the public deserved to have access to unbiased, real-time data to help form more informed viewpoints. In response, I started a website and Twitter feed called MyGridGB which made up to date and verifiable numbers on electricity available to the public. I shared pie charts of the British electricity mix on social media, blogged on where our power was coming from and tried to show all sides of our soon to be fast-evolving electricity system. The response from the public was fascinating with groups as diverse as renewable protagonists, shale gas companies and consumer protection groups looking at the same graph or tweet and drawing different but better-informed conclusions. This response, was exactly what I had set out to achieve, and in a small way, I felt that this vindicated Professor MacKay: people do well with real numbers and deserve to see them.

Since "Sustainability without the Hot Air" was published, the energy landscape has changed dramatically, with low-carbon energy generators taking their place alongside major power stations. But there remains a lack of clarity about how and when society can decarbonise the energy mix to the extent needed to prevent runaway climate change.

At the same time, I have moved forward with my career which has now taken in projects in the UK, Central Europe, East Africa, Southern Africa, Latin America, New Zealand and the Pacific, and I have seen what the new energy landscape means in hundreds of different contexts. I believe that has given me some perspective on what the future might hold for British energy.

In this book, I hope to reassess British electricity use and propose how the country can meet its carbon goals. Given that it has been nearly ten years since Prof MacKay's ground-breaking book, it is time to refresh our understanding of low-carbon energy and assess whether it is possible to meet one of the greatest challenges of today and tomorrow: electricity decarbonisation.

ABOUT THE AUTHOR

Andrew F. Crossland, CEng earned a PhD from Durham University in 2014 after spending three years researching how much value battery storage could bring to electricity networks serving two million homes. This showed how energy storage could reduce the costs of integrating solar in networks by hundreds of millions of pounds. While conducting his research, Andrew undertook interdisciplinary research into off-grid electricity in East Africa, identifying social and technical interventions to make low-carbon energy work better. The work showed how simple measures might to extend battery life from six months to five years.

After a year working on the railways in York, Andrew worked as an Energy Storage Specialist for SolarCentury in London. He helped to build a residential solar PV and battery storage offerings with for retailers which were some of the first to integrate cutting-edge anthropological research from Loughborough University. Andrew also led the development of a business case for large-scale energy storage with clients in the UK. Outside of the UK, Andrew worked on other microgrid and energy storage projects in East Africa, continuing a passion to bring energy to places underserved by fossil fuels.

As an advocate for the new energy system, Andrew was made Associate Fellow of the world leading Durham Energy Institute where he has given public lectures to academia and industry. In addition, Andrew manages the energy information service MyGridGB which charts British electricity. This was the one of the first to integrate solar data and grid carbon intensity into national electricity measurements.

Since 2018, Andrew has spent half his time in New Zealand working with a leading decarbonisation firm, Infratec. Here he is working on building up commercial and technical models to help transition utilities and businesses across the Pacific to renewable generation. He is also Director of Advance Further Energy Ltd which provides expert consultancy and techno-economic modelling for electricity generation and energy storage projects. Andrew passionately believes in low-carbon energy – human ambition is the only limit to how fast we can advance further decarbonisation.

As a result of his work inside and outside of the energy industry, Andrew was awarded the "Rising Star" award by Energy UK in 2017.

1

INTRODUCTION

Former US Vice-President Al Gore said in a 2007 address to the US Congress that

> The planet has a fever. If your baby has a fever, you go to the doctor. If the doctor says you need to intervene here, you don't say, "Well, I read a science fiction novel that told me it's not a problem."
>
> *(U.S. Government Printing Office, 2007)*

Purchasing this book, you are more likely to at least be intrigued by the link between climate change and energy. Humanity needs to address this planetary impact of energy suppliers, and in doing so, it is important to remember that these are energy supplies which have sustained, do sustain and need to continue to sustain life.

Perhaps the first time human beings took control of their own energy supplies was with the advent of fire. Fire meant that we could be warm when we wanted to be, that we could see after sunset and that we were able to eat meats which would otherwise be indigestible to our stomachs. Energy subsequently evolved at a slow pace, taking thousands of years to transition from collecting and burning firewood, to domesticating horses and harnessing wind for sailing. Despite these advances, humans were often restricted for millennia by the energy available from their local environment.

Although some long distance trade routes for food and other products had been operating for centuries, it was the industrial revolution finally tore apart the link between what the local environment could provide and human energy demand. Inventions such as the steam engine harnessed fossil fuels in a transformative way to give humans hundreds of times more energy than ever before. These machines began to demand energy beyond what local forests and mines could entirely provide. Land-, sea- and ocean-based trade routes were established which allow materials from one part of the planet to be exchanged for materials in other areas. Humans began using ships and trains to transport wood and coal hundreds of miles from where it was harvested and mined to where it was demanded in our industrialising towns and cities.

In just a few centuries, many societies transitioned from ones which depended on the horse and the tree, to ones built on distant oil wells and coal mines. As a result, energy is now available at the flick of a switch or at the pull of a petrol station pump and using fuels which are by their very nature entirely unsustainable.

The planetary impact and reach of human beings has now surpassed our ancestors. As a result, the economies of entire regions are built on providing global energy supplies, and some countries are so dependent on the industry that it must be hard for them to imagine a fossil fuel-free world. Energy lets humans enjoy life and travel in a way that was never possible before while at the same time it has been the catalyst for conflicts that have killed millions of people. Fossil fuels have transformed society in many positive, but their use is damaging our planet.

In 1859, Irish physicist John Tyndall discovered that the Earth's atmosphere had a greenhouse effect; the fundamental property which regulates the temperatures needed to sustain life. Over 150 years since that discovery, the links between human emissions of greenhouse gases and the warming planet are now beyond reasonable debate. The Paris Accords and other international treaties summarise well that international systems must change rapidly before the damage to Planet Earth becomes irreversible. Despite those accords, progress on human decarbonisation has not yet been fast enough to prevent climate change beyond what humans can contain.

In spite of the immediate need to enact the change needed, carbon-intensive industry powers processes which humans depend on. In Britain, hospitals depend on medicines which are transported thousands of miles around the world by oil burning ships, businesses require reliable and affordable energy which presently comes from fossil fuels and most homes are heated using gas or oil. There is an unspoken truth that humans need to decarbonise to make life sustainable as fossil fuels run out, while at the same time they rely on carbon-intensive processes for their survival.

Global decarbonisation is also threatened by the need to spread prosperity to new populations. Hundreds of economies are seeking to grow

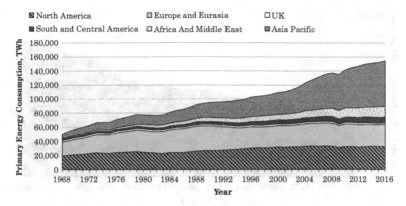

FIGURE 1.1 Global consumption of primary energy sources has grown markedly and almost consistently between 1965 and 2016. For reference, more than 385 TWh of coal and gas was used to make UK electricity in 2016, compared to a global energy consumption of over 120,000 TWh (British Petroleum (BP), 2017; DUKES, 2017a).

to provide the same opportunity that is enjoyed by the richest and most carbon-intensive countries. In order to grow their economics, energy consumption in Asia-Pacific nations increased tenfold from 1968 to 2016 (Figure 1.1), and industrialisation in these countries with their huge populations now accounts for nearly 48% of global carbon emissions. However, China with a population of nearly 1.4 billion people still produces half the carbon per capita as the United States with a population of 320 million. To achieve climate goals, it has been estimated that China alone needs to reduce the carbon intensity of its economy 60% by 2030. Combatting climate change is a global issue fraught with many social, economic and technical challenges.

As global economies have grown, anthropogenic greenhouse gas production is now higher than ever before. In 2016, annual global carbon emissions were more than two and a half times greater than in 1968 (Figure 1.2) and changes in the climate were becoming visible. In most parts of the world, carbon emissions are a direct outcome of economic output as high-income countries produce on average of forty times more carbon than a typical low-income country (Figure 1.3). The international economy is growing, and huge populations in Asia, Africa and Latin America are yet to realise their full potential for developing the large and complex energy systems seen in Europe and North America.

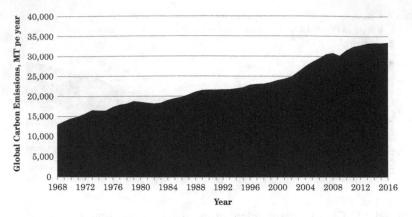

FIGURE 1.2 Total global carbon emissions increased almost every year from 1965 to 2016 as a result of growing use of fossil fuels and high-carbon industries around the world [British Petroleum (BP), 2017].

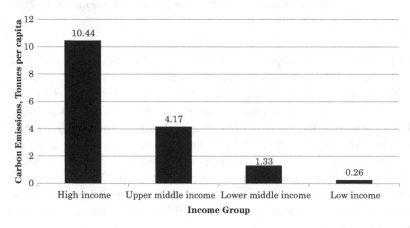

FIGURE 1.3 Average carbon emissions per capita of 217 countries and territories per income group – analysis by author using data from World Bank (2018). High-income countries have a much greater carbon emission than low-income nations; a key challenge with respect to climate change is how to reduce carbon emissions while at the same time seeking to improve quality of life for all citizens of the world.

If low-income nations developed in the same high-carbon way that their high-income counterparts have, then it is impossible for emissions to ever fall. For example, if the global population produced as much carbon as the average Chinese person, then total emissions would be more than 140%

of what they were in 2016, while if the whole world produced carbon like the average American, then emissions would be more than 350% of what they were in 2016. Yet if humanity were to lower emissions to the same level as the average person from Bangladesh, then the global greenhouse effect would be less than 15% of what it was in 2016. A key challenge in the face of climate change is finding ways of providing the energy needed to raise the quality of life, grow economies and alleviate poverty while at the same time developing the technologies and social change needed to provide that energy in a sustainable fashion.

Humans have developed numerous ways of living which are entirely damaging to the planet and to many it might seem impossible to achieve low-carbon economies when the release of greenhouse gas pervades almost all parts of international civilisation. However, positive examples of decarbonisation of some parts of the energy mix have already occurred. Portugal covered all of its electricity demand using solar, hydro and wind for four consecutive days in 2016 (Neslen, 2016) and in the same year 98.1% of Costa Rica's electricity came from renewables (Walker, 2017). In Ethiopia, over 90% of the country's electricity came from hydro and wind in 2015 (CIA, 2017), and the country is investing in wind turbines and hydroelectric power stations to increase electricity access from 27% to 90% of the population in just five years (Monks, 2017). Under some projections, China, the USA, India and the European Union (EU) all expect to run on at least 50% clean energy by 2050.

Partly as a result of decarbonisation efforts, the Global Carbon Project found that global carbon emissions practically stabilised between 2014 and 2016 (Hausfather, 2017). However, despite international agreements to reduce climate change, total greenhouse gas emissions rose again in 2018 due to a resurgence of coal use, vehicles and economic growth (Global Carbon Project, 2018). Price Waterhouse Coopers publish an index which tracks the size of the world economy alongside carbon emissions which found that the year 2015 "may be the first signs of the uncoupling of emissions from economic growth." Carbon emissions around the world are falling per unit of gross domestic product (GDP) as a result of this decoupling of emissions and the economy (Global Carbon Project, 2018) and in per capita terms were lower in 2018 than in 2008 in more developed countries (Figure 1.4). However, emissions in countries like India and China have been rising as economic growth outpaces carbon reduction per unit of GDP. This reinforces that a key dilemma is how to sufficiently further decouple greenhouse emissions and GDP in order to mitigate climate change, while expanding opportunity in both developed and developing countries.

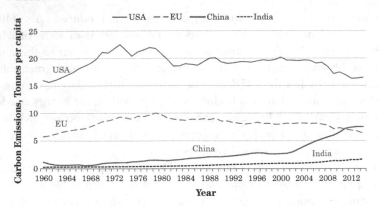

FIGURE 1.4 Carbon emissions per capita in the EU, USA, China and India. Emissions have risen in China and India, which are the two most populous nations on earth but have declined in the USA and the EU as these economies get more carbon efficient.

Carbon goals, carbon intensity and electricity generation

The impact of the energy mix on the environment is increasingly recognised in the global psyche because atmospheric carbon dioxide and other greenhouse gases from the energy sector are measurably heating up our fragile planet. Identifying a means of decarbonising the energy system, i.e. transitioning it away from processes which damage the environment, is surprisingly controversial. This is particularly true within the electricity sector.

The Intergovernmental Panel on Climate Change (IPCC) assesses the greenhouse effect of different forms of electricity generation which is summarised in Figure 1.5. This is expressed in "$gCO_2eq./kWh$" which is the greenhouse effect in grams of carbon dioxide per each unit (kWh) of electricity generated; the higher the emissions, the higher the greenhouse effect. For example, the UK Committee on Climate Change recommended that the British Government set a target emissions level in the electricity sector of below 100 $gCO_2eq./kWh$ by 2030 (House of Commons, 2016) which is a near threefold reduction from an average of over 290 $gCO_2eq./kWh$ for electricity consumed in Great Britain in 2016. This has subsequently part of a journey to net zero carbon emissions across the whole of the UK by 2050.

The IPCC numbers account for the whole lifetime greenhouse impact of a technology including carbon in production, operation and disposal. This means that wind, solar and hydroelectricity have some greenhouse impact, despite being commonly understood to be renewable electricity generators with no emissions. For example, a hydroelectric project will generate greenhouse gases in construction through setting concrete. Some solar panels are manufactured in factories where the electricity comes from coal and gas power stations which needs to be accounted in whole life carbon calculations. Similarly, offshore wind turbines are installed by ships and cranes which presently run on fossil fuels and are accessed by fossil fuel-powered boats or helicopters for maintenance. Emissions vary from project to project, and in the most extreme cases, nuclear and hydro are beyond carbon intensities in line with UK greenhouse gas emission targets[1]. That is not typical, and the key carbon values in Figure 1.5 are the median ones which are the carbon values considered for the rest of this book. Although there are some greenhouse emissions as a result of using low-carbon technologies, coal, natural gas and biomass can be far more damaging to the environment, and the median value for all of these high-carbon technologies far exceeds the UK interim goal of 100 $gCO_2eq./kWh$.

To the detriment of the climate, fossil fuels (particularly coal and natural gas) are major sources of electricity in the world (Figure 1.6). Of the low-carbon electricity sources, hydropower is more widely used than nuclear power despite the geographical constraints limiting the number of viable

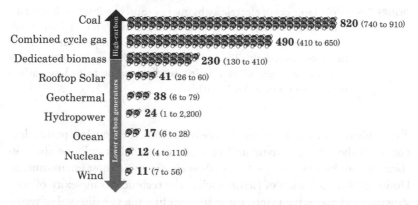

FIGURE 1.5 Median (minimum and maximum) lifecycle carbon intensity factor of different electricity generation technologies, $gCO_2eq./kWh$. Fossil fuels have by far the highest median carbon emissions even when the lifetime carbon used to make, operate and dispose of other electricity generation technologies is considered (Schlömer et al., 2014). Note that hydropower has the highest emissions of all technologies due to the high amounts of carbon that can result from construction. For the remainder of this book, the median carbon intensity is used.

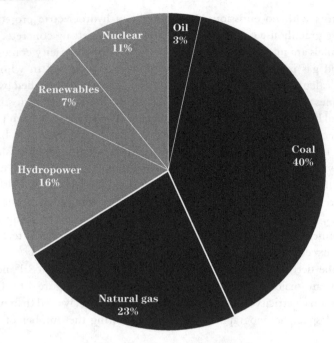

FIGURE 1.6 Global sources of electricity by energy source in 2014. Almost two thirds of global electricity come from gas, coal and oil (dark grey). Of the low-carbon electricity sources (light grey), hydropower is the most abundant. To combat climate change, a much higher percentage of global electricity needs to come from low-carbon generators (World Bank, 2017).

places where hydroelectric dams can be built. In theory, nuclear power does not suffer the same constraint and proponents of nuclear eulogise about its ability to produce predictable low-carbon electricity in most environments. However, a combination of factors such as the technical complexity of constructing and managing plants, issues surrounding the safe disposal of waste, fuel processing and high price of nuclear power generation and plant decommissioning have stunted its ability to participate in the global energy mix.

Although hydroelectricity and nuclear power have so far been unable to come close to matching, the reach of fossil fuels in the global electricity mix, the power sectors of some countries such as New Zealand and France are already highly decarbonised. In the former, more than 75% of electricity comes from renewable sources, and latter nuclear energy comprises the major part of the electricity mix. However, these are isolated examples, and to prevent further human-induced climate change, the global balance of energy needs to change from one built on fossil fuels to one built on identifiably low-carbon sources. The hope is that renewable energy sources will take over most, if not all, of the fossil fuel use in the electricity and energy sectors. That is as true in Britain as in many other countries in the world.

The British energy mix

For many reasons, Britain[2] has one of the hardest transitions of any country in the world to a sustainable energy system. The island has large energy consumption, meaning that billions of pounds of investment are needed to build new low-carbon generators; there are few viable sites for hydroelectricity owning to a lack of high mountains or large rivers; consumption of energy is mostly concentrated in the South East, far from where most electricity is produced; long summer days are great for solar, yet long winter nights mean that there can be little generation from photovoltaic panels for weeks at a time; a low geographical spread means that the whole country can have low wind output for days at a time winters can be cold, which drives highly seasonal demand for heat; and there is an acute historical and cultural dependence on fossil fuels. At the same time, high-carbon industries such as North Sea oil and gas extraction are key sectors of the economy and employ thousands of people, while untapped fuel reserves mean that the country has the potential for some energy independence for decades through fossil fuels from under the sea, coal seams and fracked/shale gas. In short, it is viable to question why and how Britain might decarbonise.

To understand the scale of the pervasiveness of fossil fuels in the British economy, it is necessary to understand the whole energy flows in the country. A breakdown of Britain's energy mix is shown in Figure 1.7. Over 75% of British energy comes from gas and petroleum which dominate the energy mix through their use in the four primary energy sectors of power, industry, transport and heating. Britain's energy consumption extends beyond the domestic sector where there is so much focus on energy saving, price reductions, insulation, green deals, etc. as nearly 40% of the final energy consumption is the oil-based fuel used in transport.

Shifting transport to electric motors and the heating system to heat pumps will require wholescale technological changes in cars and homes while at the same time will dramatically increase the size of the electricity system. The complexity of complete electrification of heat and transport is apparent when one looks more closely at all of these energy vectors. Figure 1.8 shows the British energy mix each day from 2016 to 2018. Transport fuels account

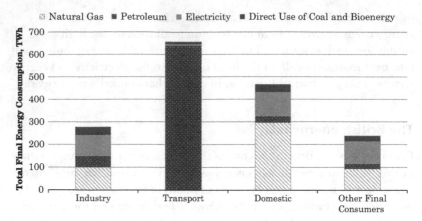

FIGURE 1.7 Division of electricity, oil, gas, coal and bioenergy between key sectors of the UK in 2017 (Department for Business, Energy & Industrial Strategy, 2018). The chart shows consumption of energy sources before efficiency losses. Efficiency is a very important part of the battle to decarbonise energy. One of the most efficient petrol engines in the world converts just 38% of the energy in fuel into work. By contrast, electric cars can be over 90% efficient. This gives perspective to how much of the energy used in high-carbon transport is not actually useful (Toyota, 2014).

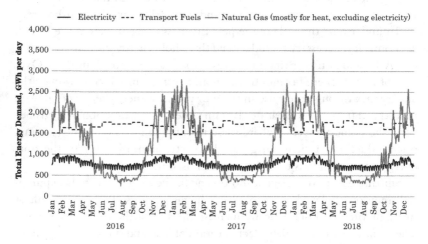

FIGURE 1.8 British electricity, gas for heat and transport energy consumption 2016–2018. Data courtesy of Dr Grant Wilson.

for more energy consumption than the entire British electricity sector while heat has the greatest demand for energy. In the summer months, less natural gas is needed for heat than the country uses for electricity, yet in the winter, gas becomes the primary energy consumer as the weather gets cold. To electrify heat requires producing an electricity system large enough to cope with energy demands in winter which are nearly three times larger than the peak electricity demand.

Complete decarbonisation requires breaking the dependence on fossil fuels by shifting to low-carbon sources and, critically, through improving efficiency. The transport sector is particularly interesting in the scale of inefficiency that pervades that energy vector. Of the petrol or diesel that is put into a car, only a fraction is actually used to propel the vehicle forwards. In 2018, Toyota reported that they had made the most efficient hybrid internal combustion engine in the world, which converted 40% of the energy in the fuel to mechanical energy (Hughes, 2018). That means in the most efficient internal combustion engine vehicles, over 60% of the energy in the fuel bought at the pump is used to move our vehicles around[3].

By contrast, battery electric vehicles can be more than 85% efficient at converting electricity from a charging cable into mechanical movement of the wheels, even if losses in the batteries and motors are considered. An equally important benefit of electrifying transport is reducing emissions of dangerous pollutants in our towns and cities which are estimated to shorten the lives of tens of thousands of people a year. However, these savings in energy and reduced tailpipe emissions are only meaningful if the energy sources used to charge car batteries or make hydrogen for fuel cells are low carbon and sustainable. British electricity does not yet meet this condition.

British electricity and how it is generated

British electricity is generated by thousands of power stations and generators spread across the country. Some generators are small such as small hydroelectric turbines in rivers which can power a kettle, while others are vast, such as West Burton coal power station, which can power more than two million homes. Electricity generators are connected together and to homes and businesses by a network of thousands of miles of cables and wires known as the grid. A single-system operator controls the electricity system, and the large power stations are switched on as needed to meet the demand for electricity. Despite many headlines proclaiming a green electricity revolution, fossil fuels were the dominant means of producing

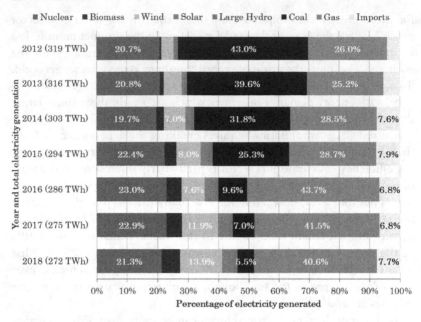

■ Nuclear   ■ Biomass   ■ Wind   ■ Solar   ■ Large Hydro   ■ Coal   ■ Gas   Imports

FIGURE 1.9   The British electricity mix between 2012 and 2018 showing a rapid decline in coal use, a switch from coal to gas, declining total electricity generation per year and the rising contribution from low-carbon electricity sources (MyGridGB, 2018). The order of values in the legend is the same as that in the chart.

British electricity in 2017. Gas was the dominant energy source in the mix, providing more than 40% of energy followed by nuclear at around 20% as illustrated in Figure 1.9 which shows how British electricity was generated between 2012 and 2018. Coal still plays a role in electricity generation, but its role is diminishing while there is a growing role for wind and solar.

It is important to appreciate the amount of energy in fossil fuels that is never converted to useful electricity, a result of the thermodynamic processes used in some power stations. A gas plant might convert up to 50% of the energy in the fuel into electricity, while a coal plant can have an efficiency as low as 35%. In both cases, energy is lost as heat and is visible leaving power station chimneys in the form of steam and smoke. In some countries, this heat is not "wasted" and is instead piped into local homes and businesses using district heating. This is the case in Lerwick, Shetland, where heat from the local oil–fired power station is circulated to local homes in the form of hot water. Understanding the scale of energy presently required to make electricity lets us appreciate how much of challenge

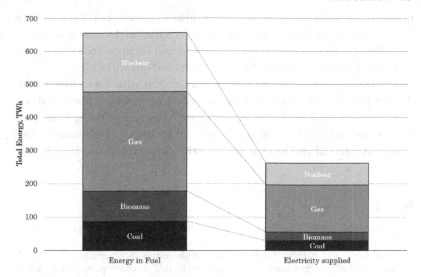

FIGURE 1.10 Energy in fuel in the supply of thermal power stations in Britain and Northern Ireland compared to the electricity actually supplied by these generators to the grid. This highlights the energy in fossil fuel feedstock which are not actually converted to useful electricity and are lost as heat (DUKES, 2017b).

it will be to transition from gas and coal. Electricity is a vast and complex part of the energy system, and in this next chapter, we will see how it has evolved to the one that is known today (Figure 1.10).

Conclusions

Decarbonisation is a necessary global objective to prevent human-induced climate change and the impacts this can have on livelihoods, the environment and our health. Making electricity systems sustainable is a global problem which requires all people and nations to act. As such the UK has a moral obligation to decarbonise, whatever international colleagues choose to do. Although doing so is an environmental responsibility, it is important to remember that the energy sector enables homes, transport, entertainment and business. To be accepted by the public, sustainability must not try to change the outputs from our energy system; instead, it should change the inputs and processes in energy in order to improve the quality of life for all.

Almost all of British energy is sourced from fossil fuels which are in limited supply and high in carbon. As a result, British energy gets less sustainable every minute of every day as fossil fuels get ever closer to running out. Until recently, local industries provided the majority of the fuel used in the energy sector, yet as of 2016, the majority of gas, coal and petroleum are imported meaning that Britain is dependent on others for its economy and quality of life.

Electricity is a key sector of the energy system which will have to undergo rapid and fundamental change in order to protect our planet and way of life. As we shall see in the next chapter, change in electricity supplies have historically been a regular occurrence in Britain. Understating that change is essential and hints at how an alternative future for British electricity might occur.

Notes

1 Where the target is for average carbon emissions in the power sector to fall below 100 gCO_2eq./kWh by 2030 as part of a journey to net zero emissions.
2 This book has a heavy focus on Britain (England, Scotland and Wales), although the concepts used are entirely applicable to other parts of the world. In some sections, the United Kingdom as a whole is referred to, in which case a narrative pertaining to the whole of Britain plus Northern Ireland is being discussed.
3 Before any losses associated with friction, air resistance, braking, etc. are considered.

References

British Petroleum (BP), 2017. *BP Statistical Review of World Energy*, s.l.: s.n.
CIA, 2017. *CIA World Factbook: Ethiopia.* [Online] Available at: www.cia.gov/library/publications/the-world-factbook/geos/et.html [Accessed 18 08 2018].
Department for Business, Energy & Industrial Strategy, 2018. *Energy Flow Chart 2017.* [Online] Available at: www.gov.uk/government/statistics/energy-flow-chart-2017 [Accessed 15 08 2018].
DUKES, 2017a. *5.3. Fuel Used in Generation,* London: s.n.
DUKES, 2017b. *Table 5.6 Electricity Fuel Use, Generation and Supply.* [Online] Available at: www.gov.uk/government/statistics/electricity-chapter-5-digest-of-united-kingdom-energy-statistics-dukes [Accessed 17 06 2018].
Global Carbon Project, 2018. *2018 Carbon Budget and Trends,* s.l.: s.n. [Online] Available at: www.globalcarbonproject.org/carbonbudget [Accessed 08 12 2018].
Hausfather, Z., 2017. *Analysis: Global CO2 Emissions Set to Rise 2% in 2017 After Three-Year "plateau."* [Online] Available at: www.carbonbrief.org/analysis-global-co2-emissions-set-to-rise-2-percent-in-2017-following-three-year-plateau [Accessed 05 12 2017].
House of Commons, 2016. *House of Commons, Energy and Climate Change Committee. Setting the Fifth Carbon Budget. Fifth Report of Session 2015–16.* [Online] Available at: https://publications.parliament.uk/pa/cm201516/cmselect/cmenergy/659/659.pdf [Accessed 9 05 2018].

Hughes, J., 2018. *Toyota Develops World's Most Thermally Efficient 2.0-Liter Engine.* [Online] Available at: www.thedrive.com/tech/18919/toyota-develops-worlds-most-thermally-efficient-2-0-liter-engine [Accessed 18 08 2018].

Monks, K., 2017. *Riders on the Storm: Ethiopia Bids to Become Wind Capital of Africa.* [Online] Available at: http://edition.cnn.com/2016/12/20/africa/ethiopia-wind-power/index.html [Accessed 18 05 2017].

MyGridGB, 2018. *MyGridGB.* [Online] Available at: www.mygridgb.co.uk [Accessed 17 06 2018].

Neslen, A., 2016. *Portugal Runs for Four Days Straight on Renewable Energy Alone.* [Online] Available at: www.theguardian.com/environment/2016/may/18/portugal-runs-for-four-days-straight-on-renewable-energy-alone [Accessed 18 05 2017].

Schlömer, S., Bruckner, T., Fulton, L., Hertwich, E., McKinnon, A., Perczyk, D., Roy, J., Schaeffer, R., Sims, R., Smith, R., Wiser, R., 2014. *Annex III: Technology-specific Cost and Performance Parameters. In: Climate Change 2014: Mitigation of Climate Change. Contribution of Working Group III to the Fifth Assessment Report of the Intergovernmental Panel on Climate Change,* Cambridge, UK and New York: Cambridge University Press.

Toyota, 2014. *Toyota Develops Engines with Improved Thermal, Fuel Efficiency.* [Online] Available at: https://newsroom.toyota.co.jp/en/detail/1693527 [Accessed 21 10 2019].

U.S. Government Printing Office, 2007. *House Hearing, 110 Congress, Perspectives on Climate Change March 21, 2007.* [Online] Available at: www.govinfo.gov/content/pkg/CHRG-110hhrg37579/html/CHRG-110hhrg37579.htm [Accessed 16 04 2019].

Walker, P., 2017. *Costa Rica's Electricity was produced almost entirely from renewable sources in 2016.* [Online] Available at: www.independent.co.uk/environment/costa-rica-renewable-energy-electricity-production-2016-climate-change-fossil-fuels-global-warming-a7505341.html [Accessed 18 05 2017].

World Bank, 2017. *3.7 World Development Indicators: Electricity Production, Sources, and Access.* [Online] Available at: http://wdi.worldbank.org/table/3.7# [Accessed 30 01 2019].

World Bank, 2018. *World Bank Open Data.* [Online] Available at: https://data.worldbank.org/ [Accessed 07 07 2018].

2

BRITAIN'S EVOLVING AND DECARBONISING ELECTRICITY SYSTEM WHERE CHANGE IS NOTHING NEW

England, Scotland and Wales have a synchronised electricity system of generators, consumers and cables, which is one of the most advanced in the world. Using this power system to keep the lights on is a complex task that is undertaken with a continually changing range of technologies, regulations and consumer demands. On a regular basis, new power stations are being built while others are being decommissioned, and every flick of a switch means that the demand for electricity changes thousands of times every second. The amount of power being generated also varies because renewable energy sources like solar and wind are constantly changing their output as the weather changes, while outages of coal, gas and nuclear power stations are more regular than the average consumer might appreciate. The electricity sector is a dynamic industry which is continually adapting to changing demand and technology.

The first electricity systems in Britain were a patchwork of small networks serving individual towns and cities. In the 1920s and 1930s, these isolated systems were slowly connected together to form a unified electricity grid which extended from the North of Scotland to the South of England. The resulting *national grid* is a continually evolving system of energy flows moving through power cables, managed through markets and backed by international trade routes importing fuel. Understanding how British electricity evolved over the past forty years (Figure 2.1) provides a unique lens into what a decarbonised energy system might look like in the future.

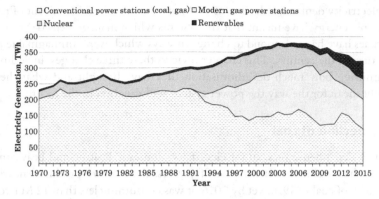

FIGURE 2.1 The UK electricity supply mix has evolved from one almost entirely made up of coal in the 1970s to one with a mix of gas, nuclear and renewables by 2015. Change is not new to British electricity (DUKES, 2017).

Between 1970 and around 1992, Britain had a growing electricity demand which was met by a mix of coal, gas and nuclear power stations. This aligned well with a country with a strong coal industry and being the first to commercialise the use of nuclear power for the peaceful purpose of generating electricity. In the late 1980s, there was a debate about how to better use gas resources from the North Sea — should Britain sell it abroad and create a sovereign fund like oil-producing nations like Norway or should Britain burn that gas to make electricity? Britain chose the latter, and from 1992 to 2004, as electricity demand grew more rapidly than before, gas power stations were constructed all over the country. Some conventional power stations using steam were decommissioned in this time, but Britain still increased the use of fossil fuels in a *dash for gas*. Towards the end of the millennium, fears of climate change were beginning to get political recognition. The Kyoto Protocol was signed in 1992, environmental laws were being enacted on a national and international level and there was a realisation that the gas reserves in the North Sea would at some stage be depleted. From 2004 onwards, electricity consumption fell as a result of economic change and increasing efficiency in appliances, while from 2010 onwards, the advent of wind and solar power stations along with this demand reduction meant that fossil fuels made a smaller contribution to the electricity mix than ever before.

The rate of change in British electricity over the forty years from 1970 to 2010 was relatively modest when compared to three major and coincident changes that occurred between 2011 and 2017. Over this time, a reduction in electricity demand, a decline of coal power stations and the advent of renewable energy have meant that the systems which managed electricity for decades have had to respond to change in ways which were unimaginable at the turn of the century. This chapter reviews these three changes, including examining how much decarbonisation they have brought and what they might mean for the way the power system will function in the future.

The decline of coal

British dependence on coal for electricity, domestic heating and heavy industry has declined over the past fifty years; the country produced nearly 150 MT of coal in 1970, yet by 2016, it was consuming less than 12 MT for electricity generation in 2016 (Department of Business Energy and Industrial Strategy, 2018). The decline in coal for electricity has been particularly dramatic; in 2012, coal power stations produced more than 40% of British electricity, but by 2017, those same plants met less than 7% of national demand. In 2016, there were over 200 hours where no coal was being used to generate electricity, increasing to over 1,000 hours of coal-free electricity in 2018. The impacts of the coal shutdown are visible in all seasons; coal transitioned from providing 14 TWh of electricity a month to less than 2 TWh in winter when electricity demand was highest (Figure 2.2).

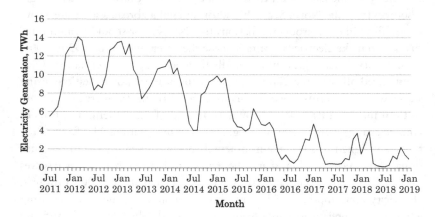

FIGURE 2.2 Electricity generation from coal every hour from May 2011 to December 2017 (Gridwatch, 2018; MyGridGB, 2018). The amount of electricity generated from coal has fallen rapidly over this time, but as of 2018, the technology is still used to meet high electricity demands during cold winter days when renewable output was low.

Coal has huge political significance in the UK as the fuel which drove the industrial revolution and the Empire yet the fuel is now widely considered to represent the harsh reality of unsustainability (both in a planetary sense and in an economic sense); the collapse of the British coal industry led to huge industrial strife in the 1970s and 1980s which many areas of the country are yet to recover. The decline of coal also represents a significant turning point in the psyche of Great Britain in terms of how electricity is generated in response to carbon emission targets. From 2012 to 2018, there was not only a steady decline in total fossil fuel use but also a dash for gas (Figure 2.3) which has mistakenly been lauded as a good thing for the planet. Gas is far less carbon intensive than coal, yet it is important to remember, as identified by the Intergovernmental Panel on Climate Change (IPCC), gas cannot be classified as a low-carbon means of generating electricity. By replacing one fossil fuel with another, it might be argued that Britain simply demonstrated an ability to switch between high-carbon electricity generators. This is a decision which has had geopolitical and economic consequences for the country. In the 1970s, the majority of coal was produced locally from British mines, yet in 2017, much of Britain's gas and coal were imported. For example, gas was imported through pipelines with Belgium, the Netherlands and Norway and as liquefied natural gas from countries like Algeria, Qatar, Russia and the USA (Department for Business, Energy & Industrial Strategy, 2018a, b). The resulting increase in international trading for the fuel needed for electricity and heating impacts energy security. If Britain were to completely divorce from coal as a source

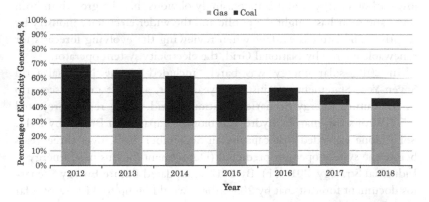

FIGURE 2.3 Electricity from gas and coal per year 2012–2017 showing the decline in coal, growth in proportional gas use and a decline in the fraction of fossil fuel generation in British electricity (MyGridGB, 2018).

of electricity and shift to gas, then the price of that electricity will be much more exposed to the market price of gas. Britain can no longer switch from gas to coal power stations to respond to rises in the gas price or falls in the price of coal and so it could be argued that this leaves the country with less ability to play international energy markets to keep prices stable. This might become a significant problem if there is an international dash from coal to gas in a global effort to reduce carbon emissions from electricity.

A rapid decline in coal is only good if it is done so in a sustainable and commercially viable manner. Thankfully, in addition to a rise in gas use, there have also been two other changes to electricity which can be considered to be low-carbon and potentially a good thing for the economy: falling demand and more renewable electricity generation.

Growing renewable energy

It is remarkable to think that when I started my PhD in 2011, wind turbines and solar panels were much more expensive than conventional power stations, widely seen as technically unviable and were usually considered to be little more than curiosities of advanced research. However, in just six years, renewable energy generators became a widespread feature of our landscape with solar panels on thousands of homes and large offshore wind farms around the British Isles. The rise of renewable energy means that in just six years, solar and wind generators have grown to provide over 18% of British electricity. The rate at which renewables have entered the national consciousness is evident through newspaper headlines citing records in low-carbon energy, and it is increasingly obvious that the growth in both wind and solar has caught the public and the wider electricity industry by surprise. The latter is evident when reviewing the evolving forecasts for renewable energy by National Grid, the electricity system operator.

In 2010, solar energy was barely included in the National Grid Seven-Year Statement, yet just two years later, a "gone green" scenario in the Future Energy Scenarios (National Grid, 2012) report projected 8.5 GW of solar, marine, hydro and biomass capacity in Britain by 2018. Solar alone exceeded that capacity in September 2015, over three years before the system operator predicted (Department for Business, Energy & Industrial Strategy, 2018a, b). By 2017, an updated Future Energy Scenarios document forecast that by 2050, there would be up to 44 GW of solar photovoltaic (PV) generation, recognising that "Technical progress and significant cost reductions in technologies, such as storage and solar panels, have driven major change in a short space of time."

The decline in coal has been concurrent in part with increasing numbers of wind turbines in Britain. The 2010 Future Energy Scenarios predicted that 13.3 GW of onshore and offshore wind capacity would be installed in Great Britain by 2018, yet across both Britain and Northern Ireland, that figure was reached in 2015 (DUKES, 2017). At the end of 2017, Britain had 30 GW of wind and solar generation installed, a capacity which far exceeded any National Grid forecast. To provide some context of the scale of renewable generation in Britain, if all of these turbines and panels were producing full output on a sunny and windy day, they could meet 50% of British electricity demand (National Grid, 2018). This is quite impressive for technologies which barely featured in the minds of one of the most important companies in British electricity just seven years previously (Figure 2.4).

Wind and solar were so successful for a number of concurrent reasons. Strong Government backed subsidies provided stable economic returns to investors in both the UK and other countries around the world, the emergence of which provided a predictable and bankable market for wind turbine and solar panel manufacturers. As a result, they could invest in the factories and scientific research needed to bring prices of both technologies down; as a result the cost of wind generated electricity in 2018 was almost a fifth of what it was in 2010 according to statistics from Bloomberg (2018). Prices fell so fast that wind and solar plants became increasingly profitable for investors, and for many years, government and industry were locked in a battle where subsidies were regularly dropped to try to control the returns that investors were making. By the end of 2017, both wind and solar electricity generators

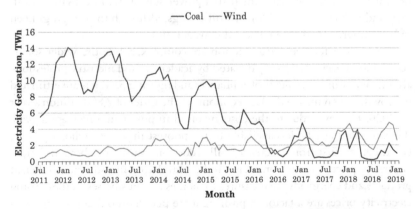

FIGURE 2.4 Electricity generated by coal and wind in Britain between July 2011 and January 2019 showing when wind began to overtake coal in the British electricity mix (Gridwatch, 2018; MyGridGB, 2018).

were beginning to reach a point in some parts of the world where they could generate at a lower price than conventional power stations without subsidy. There is evidence that that was also true in the UK.

Coal use in Britain has been replaced by both gas and renewables. Unlike gas, alternatives like solar and wind are a low-carbon and sustainable means of generating electricity. The final change in the electricity system since 2011 is also arguably low-carbon and one that is rarely mentioned: a fall in the amount of electricity that Britain consumes.

Falling demand for electricity from major power stations

From 1970 onwards, electricity supplied to the UK steadily rose to a peak in 2005 through a growing services sector and increasing demands in homes. Since then, consumption of electricity from Britain's major power stations has fallen due to a combination of factors, from a changing economy, rising prices, growth of small-scale renewables and the increasing efficiency of electrical appliances and lighting. The decline in electricity use is mirrored in the entire energy sector, where it is estimated that consumption fell by 14% on both an actual and temperature corrected basis despite a rise in population (Vella, 2017).

From the perspective of the carbon intensity of electricity, falling consumption usually means a reduction in the amount of energy that needs to be produced by high-carbon power stations. Demand reduction is particularly effective as a means of reducing fossil fuels because of the way that the electricity system works. Fossil fuels are usually one of the last power stations to be used to meet electricity demand, so by reducing consumption, it is more likely that a fossil fuel power station will be switched off. Demand reduction is good for carbon savings, although these savings need to be tempered by the wider social considerations.

The reasons for demand reduction are complex, while the wider socioeconomic impacts are heavily debated by academics and industry professionals. Efficiency improvements of appliances and lighting products are estimated to provide net savings to the UK economy of the order of £850 million a year by 2020. However, due to the costs involved in purchasing new appliances, these are naturally be less available to the poorest in society who, arguable, could benefit first from electricity bill reduction. Similarly, above inflation, electricity price rises are considered to be stimulating investment in efficient products and energy saving measures by homes and businesses. If these rising electricity prices are a factor in pushing more people into fuel poverty, then one of the basic criteria of a socially responsible power system (providing the energy needed for a healthy and high quality life) is not being met.

The story of carbon in British electricity

On an hourly basis, the carbon as a result of electricity generation constantly changes as the electricity mix responds to changes in system demand, power station reliability, dispatch of fossil fuel generators and the output of variable renewables. For example, in November 2016, the carbon intensity of British electricity peaked above 450 $gCO_2eq./kWh$ when cold weather led to high demand, and there were large numbers of coal power stations generating electricity. However, just six months later in May 2017, carbon intensity was briefly below 180 $gCO_2eq./kWh$ when solar output was high. Social media often draws attention to the peaks and troughs of our greenhouse gas production. However, it is worth remembering that low-carbon emissions on a sunny day in the summer can be as misrepresentative of long-term decarbonisation treads as high-carbon emissions in the winter if coal plants are online. The underlying changes in carbon emissions which occur over a period of years are much more relevant.

In 1970, the majority of British electricity was produced in high-carbon coal power stations so that electricity had a very high-carbon intensity. This was followed by a shift to more low-carbon electricity generation such as nuclear and gas, yet because electricity demand rose between 1970 and 2006, total carbon emissions marginally increased, peaking above 200 MT per year. Over this time, diversifying the electricity mix to more low-carbon sources merely tempered the increased electricity demand resulting in no net saving in total greenhouse gases. Carbon emissions only began to drop with the combined effects of demand reduction, rising renewables and falling coal use from 2012 onwards (Figure 2.5). In 2012, a carbon intensity of 500 $gCO_2eq./kWh$ was a regular occurrence, while in 2017, the total carbon intensity fell below 280 $gCO_2eq./kWh$ over the year, which represents a near 50% reduction in carbon intensity in just five years. In fact, total carbon emissions in Britain in 2017 were below what they were in 1970.

As summarised in Figure 2.6, three major changes in the electricity have caused these carbon savings. Diversifying from coal power stations has had the largest effect by abating more than 35 MT of carbon in 2016. The actual carbon saving from demand reduction is somewhere in the range 18–31 MT and the carbon saving from renewables in the range 13–22 MT.[2] Despite the fact that total carbon emissions from the electricity sector have fallen, Britain is still far in exceedance of a 2030 carbon interim goal set by the Committee on Climate Change. Britain is missing that target because

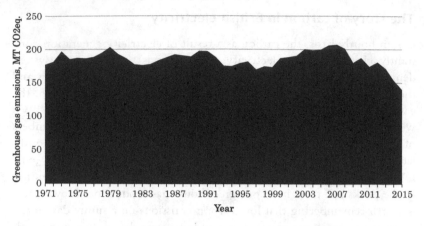

FIGURE 2.5 Total greenhouse gas emissions from British electricity have only recently started to decline. This is because a decline in the carbon intensity[1] of electricity before 2012 was insufficient to offset by increasing electricity consumption [analysis by author using data from DUKES (2017)].

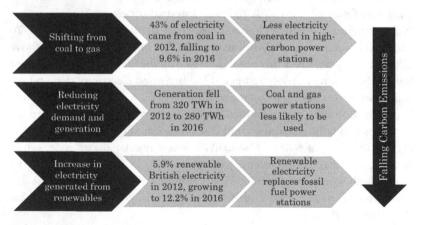

FIGURE 2.6 Impact of demand reduction, a shift from coal and growth of renewables on decarbonisation.

a significant proportion of electricity in 2018 still comes from high-carbon sources. As decarbonisation continues, a strategy of switching off coal power stations has only limited future effect since Britain has few operational coal power stations left. As a result, policy must increasingly focus on lowering demand (which itself has a finite impact before the function of the power system itself is diminished) and increasing the role of low-carbon electricity generation.

Coal use has declined so much that focus must move onto reducing the use of natural gas which forms the bulk of the electricity sectors carbon emissions. I was once quoted in the *Daily Mail* as saying that "if we continue to run the system with this much gas we will miss our carbon targets" (Press Association, 2017). That is as true today as when I said it in 2017.

Conclusions

Change is normal in British electricity and something which must be applauded, embraced and encouraged if the country is ever to have a truly sustainable energy system. Between 2012 and 2017, Britain transformed itself from one of the highest carbon power systems in the world, with vast amounts of coal being continually burned in power stations, to one less dependent on the world's most carbon-intensive fossil fuel. The country is seen as a global leader in low-carbon electricity with renewable generators installed everywhere from on the roofs of homes to the seas surrounding the country. Energy efficiency and demand reduction have also had an impact by reducing the need for fossil fuels to produce electricity.

Despite all of these changes, Britain in 2017 was still a long way off achieving climate targets because low-carbon generators still provided less than 50% of the country's electricity supply (Figure 2.7).[3] Achieving climate

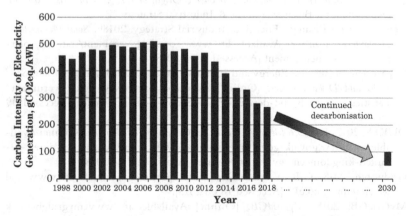

FIGURE 2.7 Carbon intensity of British electricity from 1990 to 2018 showing the progress that needs to be made to bring emissions below 100 gCO$_2$eq./kwh by 2030 (MyGridGB, 2018).

goals is non-negotiable if irreversible climate change is to be avoided, and it is widely recognised that decarbonisation efforts must continue, and gas is the new villain of the decarbonisation agenda. Before establishing whether it is possible to shift away from gas, the next chapter will assess what tools the country might have for doing so. In this, a low-carbon toolkit is introduced which is what utilities, policymakers and engineers might use to build a decarbonised energy future.

Notes

1 The amount of carbon produced per unit of electricity.
2 Approximating the carbon saving from demand reduction and renewable is more challenging as it is impossible to know what proportion of gas and coal use they have reduced. This estimate assumes a low margin emission factor equivalent to the carbon intensity gas power stations and a high emission factor to be the carbon intensity of coal power stations.
3 With 50% of electricity eventually coming from low-carbon sources in 2018.

References

Bloomberg, 2018. *Coal Is Being Squeezed Out of Power by Cheap Renewables.* [Online] Available at: www.bloomberg.com/news/articles/2018-06-19/coal-is-being-squeezed-out-of-power-industry-by-cheap-renewables [Accessed 20 01 2019].

Department for Business, Energy & Industrial Strategy, 2018a. *Energy Trends Table 4.4 Supplementary Information on the Origin of UK Gas Imports,* London: Department for Business, Energy & Industrial Strategy.

Department for Business, Energy & Industrial Strategy, 2018b. *Solar Photovoltaics Deployment.* [Online] Available at: www.gov.uk/government/statistics/solar-photovoltaics-deployment [Accessed 11 08 2018].

Department of Business Energy and Industrial Strategy, 2018. *Energy Trends: Solid Fuels and Derived Gases.* [Online] Available at: www.gov.uk/government/statistics/solid-fuels-and-derived-gases-section-2-energy-trends [Accessed 19 08 2018].

DUKES, 2017. *Table 5.6 Electricity Fuel Use, Generation and Supply.* [Online] Available at: www.gov.uk/government/statistics/electricity-chapter-5-digest-of-united-kingdom-energy-statistics-dukes [Accessed 17 06 2018].

Gridwatch, 2018. *G.B. National Grid Status.* [Online] Available at: www.gridwatch.templar.co.uk/ [Accessed 18 08 2018].

MyGridGB, 2018. *MyGridGB.* [Online] Available at: www.mygridgb.co.uk [Accessed 17 06 2018].

National Grid, 2012. *Future Energy Scenarios 2012,* Warwick: National Grid.

National Grid, 2018. *UK Future Energy Scenarios,* Warwick: National Grid.

Press Association, 2017. *Rise of Renewables as Wind Farms Power Past Coal Plants.* [Online] Available at: https://www.dailymail.co.uk/wires/pa/article-5217161/ Rise-renewables-wind-farms-power-past-coal-plants.html [Accessed 23 01 2018].

Vella, H., 2017. *What's Driving the Fall in UK Energy Demand and Can It Last?* [Online] Available at: www.power-technology.com/features/featurewhats-driving-the-fall-in-uk-energy-demand-and-can-it-last-5819050/ [Accessed 03 02 2019].

3
DESIGNING A LOW-CARBON ELECTRICITY SYSTEM

Kofi Annan once said that "*Shifting towards low-carbon energy systems can avert climate catastrophe while creating new opportunities for investment, growth, and employment*" (2015). However, how do you even begin to design an energy system which can achieve all of those objectives?

You would be hard pressed to find a sector in the economy that does not need to reduce its greenhouse gas emissions in order for climate targets to be met. Householders need to reduce their demand, agriculture needs to reduce methane emissions, industries need to find decarbonised manufacturing methods, designers are needed to produce pollutant-free vehicles and the electricity system needs to transform into one almost entirely free of fossil fuels. This near universal behavioural, economic and technical change is often seen as a major challenge to combatting climate change because it requires all to forget about the competing demands on time and resources and do something to make a difference towards climate change. Unfortunately, the world does not always appear unilaterally intent on making those changes.

Decarbonising is seen as a threat by entrenched interests who are trying to protect a status quo that they benefit from. In the 1960s, the tobacco industry freely promoted fictional health benefits from cigarettes and funded dubious scientific research to support their threatened businesses. It took decades to dispel the myths that the tobacco industry supported. Today's energy industries are being accused of behaving in a similar way, with the aim of prolonging the fossil fuel industry for as long as possible.

We have already seen that one of the three pillars of Britain's decarbonisation strategy is switching off coal power stations and partially replacing them with gas and a really clear example of this strategy is a line often stated by anti-low-carbon interest groups: "we cannot run an energy system without gas."

A strategy that might be used to maintain a British fossil fuel industry for the medium term is presented in Figure 3.1. This is a somewhat cynical presentation of how an entrenched energy industry might behave to protect their interests, but one which demonstrates how the politics of energy,

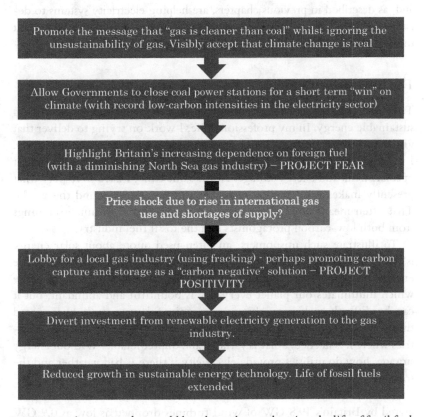

FIGURE 3.1   A strategy that could be adopted to prolonging the life of fossil fuel use in a country. Misinformation and uncertainty are tools that can be used to prevent change in an established energy system. The public and policymakers deserve to know the facts from all perspectives in order to form and continually challenge opinions and behaviours.

from both pro and anti-low-carbon power advocates, should usually be treated with a degree of scepticism and care.

Thankfully, in addition to expanding operations in gas, established energy companies are increasingly investing in alternative energy, with BP, Shell, Total, Danish Oil and Natural Gas, etc. (all major oil companies) now rebranding as global energy organisations as opposed to be fossil fuel-dependent businesses. This divestment is clear evidence that new energy is a threat to the establishment, much like video streaming over the internet has been to some traditional media companies.

Alternative and low-carbon energy sources are becoming more prevalent and, as described in previous chapters, are helping electricity systems to decarbonise. However, to get most value from alternative energy, it is important to understand what makes up a viable toolkit of methods for decarbonising.

The low-carbon toolkit

The dream of climate change activists is a society which depends only on sustainable energy. In my professional life, I work on trying to deliver that vision to homes, businesses and whole islands. To do my job effectively, I need to understand the respective strengths and weaknesses of a toolkit of solar, wind and energy storage. At the same time, I have to respect what presently makes fossil fuels so valuable and prevalent around the world. That often means coming up against misnomers and misunderstandings from both low-carbon protagonists and the fossil fuel industry.

To illustrate such misnomers, an often-used quote about solar energy is that "every 24 hours, enough sunlight touches the Earth to provide the energy the entire planet needs for 24 years." It is true that the solar resource which illuminates our planet every day is bountiful and abundant, but it can be difficult to harness that in a meaningful way; sunlight is spread thinly across the planet and far from many of the towns, cities and industries where it is needed. Solar is variable and unpredictable. Without energy storage, how do utilities provide electricity if there is bad weather and the solar output drops below what is needed?

Wind is equally variable: in the month of June 2018, the electricity generated by Britain's 19.8 GW of wind turbines dropped as low as 0.9 GW when wind speeds were low. Hydroelectricity can be much less variable than wind or solar, but when shortages in hydro generation do occur, they can last for months rather than weeks for wind or days for solar. In October 2017, an ongoing drought led to hydropower shortages in Brazil and a 43% rise in power prices.

To overcome such issues, it is necessary to install a range of complimentary electricity generation technologies which help support each other when their output is low. This is called an *Electricity Mix*, and creating that mix is the most important item in a low-carbon toolkit.

Chase an energy mix

An electricity mix means having a diverse mix of methods for generating power in order to reduce exposure to weaknesses on one particular technology. I strongly believe that having a diverse energy mix is important for all consumers whether they wish to remain dependent on fossil fuels or to decarbonise. Countries with a high dependence on coal power are exposed to the changing prices of the international coal market, while countries with high dependence on the same type of nuclear power stations will face issues if a flaw within those plants is found and all need to be repaired. As an example, four French nuclear plants were shut down by a common issue during a heat wave in August 2018, which raised national electricity prices (Reuters, 2018).

In New Zealand, being green is part of the national identity and the country's beautiful environment which draws tourists from around the world. The nation is a green champion with over 70% of electricity being generated from low-carbon sources;[1] however, New Zealand is also a good example of a country which has all of the right intentions on decarbonisation but is missing out on the benefits of a truly diverse energy mix.

Because of the dominance of hydro, the focus of the entire electricity system is on dry year insurance for when there is not enough water in the hydroelectricity dams. So critical is the hydropower resource that New Zealand's grid operator, Transpower, tracks the risk of hydro shortage on their website. Reliance on a single set of cables running under The Cook Strait linking the two islands also means that hydroelectric generation on the South Island can be taken offline through the loss of the subsea connection. The potential for loss of hydro production during a dry year used by some to support the retention of gas and coal power stations as an insurance.

Rather than accepting their present electricity system as the status quo, New Zealand could increase the diversification of their electricity mix with new low-carbon generation. Mixing hydropower with much greater levels of solar and wind would mean that the country would only call on gas power stations when it is consistently dark, there is no wind and there is a drought, an occurrence which is highly improbable in a large country like New Zealand. Such a strategy would not mean covering the environment with solar panels or wind turbines and would make the country one of the few industrialised economies to generate electricity almost entirely from renewable power stations using a viable low-carbon electricity mix.

Such a scenario is shown in Figure 3.2. In the upper figure, the electricity mix of the country is shown at a fifteen-minute resolution for two weeks in

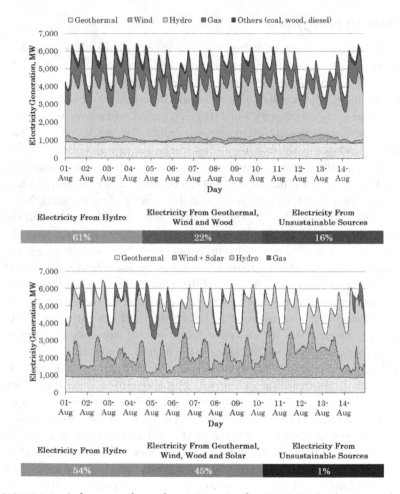

FIGURE 3.2 A lower carbon electricity mix for New Zealand using solar and wind alongside hydro to displace fossil fuels is very easy to achieve. The top chart shows how electricity was generated in the first week of August 2018. The bottom chart shows electricity could have been almost completely decarbonised in New Zealand through solar and wind power which can be generated at a lower cost (analysis by author).

August 2018 as it was actually generated. Over this time, New Zealand actually generated 15% of its electricity from gas power stations by running them 24 hours a day. In the lower chart, weather data has been used to simulate the introduction of solar and wind into the system. Adding in a modest amount of solar and wind and using it as an alternative to gas leads to a very different energy mix with fossil fuels generating just 2% of the electricity. The gas is only used as a backup on days where there is no wind and little sun.

New Zealand's present approach to electricity is like turning up to a football game with six strikers (hydro) to score goals but retaining five keepers (fossil fuels) as a defence: a strategy never used in football. With a low-carbon electricity mix, it would have a team of strikers, midfielders and defenders (solar, wind and hydro) who work together to create goal scoring opportunities and retain a single keeper (gas) to prevent the opposition scoring if the defence fails. This can leave New Zealand exposed to periods of very high prices (Figure 3.3).

Wind turbines, hydro and solar are variable resources of electricity generation, and their effectiveness will always be affected to the weather conditions. One of the most widely quoted arguments against renewables is

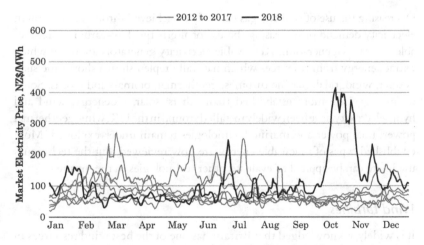

FIGURE 3.3 In 2018, power prices in New Zealand hit record levels. In the press, this was blamed on low lake levels due to a dry summer and gas shortages. I would argue that a lack of diversity in New Zealand's electricity mix left them vulnerable to the volume of rainfall. Low-carbon diversity could come from solar, wind working alongside hydroelectricity and geothermal; as a year of no rain, no sun and no wind is extremely unlikely (Electricity Authority, 2018).

that they can never be bulk providers of energy because there will always be some dates a year when there is insufficient electricity generation from them to meet demand. Examples include cold winter days when demand rises and the air is still and/or the skies are cloudy.

It is important to be fully aware that decarbonised energy production (whatever the form) has a variable output; some days are sunnier than others. This is true on the Pacific Islands, where solar and battery storage provide up to 95% of the electricity supply. Although solar is abundant in this region, there will always be some days of prolonged poor weather where there will be a shortage of sunlight. To provide power on days of low solar irradiance, it is typical to use a diesel generator. It might seem counterintuitive to retain a diesel generator on a renewable energy plant, but the islands are transitioned from using imported diesel fuel all of the time to one using diesel as a backup only. Achieving 95% of electricity from renewable sources is a huge leap forward and one that uses a diverse energy mix to manage the technical strengths and weaknesses of the available tools. So how might a country like Britain grow its own low-carbon energy mix to reduce coal and gas use?

Renewable electricity

Decreasing the use of coal and gas can only be achieved through reduction in electricity demand or increasing the use of alternative low-carbon or renewable electricity generation. Renewable electricity generators are those which extract energy from resources which naturally replenish in a short time such as solar, wind, tidal, marine turbines, geothermal, biomass and biogas. Some technologies are more established than others: solar, bioenergy, wind and hydroelectricity have been widely installed around in the UK, while geothermal power, tidal power and marine technologies remain under-exploited. More established types of renewable energy are now reviewed, but the techniques used here can be applied to evaluate other technologies in the future.

Wind turbines

It is widely acknowledged that Britain has one of the best wind resources in the world as a result of its position on the north-eastern edge of the windy Atlantic Ocean. As a result of various subsidy and regulatory regimes, the wind industry in the UK almost doubled between 2012 and 2016 and wind turbines generated 50% of the renewable electricity in Britain in 2017 (DUKES, 2018). This is a staggering growth for a relatively modern technology, but as of 2015, wind turbines are not a major part of the global electricity system. That might change as a result of the falling cost of producing electricity from wind. Analysis of global wind power prices by Bloomberg found delivered costs of wind energy in the UK fell from around $90/MWh in 2016 to below $40/MWh in 2022 (Bloomberg New Energy Finance, 2018a). These price drops have been driven by improved manufacturing, more efficient turbines, being able to install turbines in harsher environments with stronger and more regular wind and by increasing the size of turbines to improve economies of scale. The largest turbines in development in 2018 were approaching the height of the Eiffel Tower.

Although costs have fallen, it would be senseless for manufactures and developers to invest in wind projects without knowing that secure contracts are available for their products due to the high upfront capital requirement and relatively low operational costs; once a turbine is built, then most of the investment is locked up in machinery. Recognising that, contracts by governments and utilities around the world now provide guaranteed income for wind turbines over large parts of their operational life, and by doing so, wind developers are given the security needed to invest large sums in wind farm projects with low costs of finance. Lowering costs of finance and reducing investment risk can have as much of an impact in reducing the cost of electricity from wind turbines as improvements in engineering or falling technology costs. Bankable projects with secure long-term contracts can be less risky for investors, will attract cheaper financing and so electricity generated can be sold at a lower price.

Historically, wind may have appeared to be a heavily subsidised form of electricity, but that investment has now resulted in wind farms being contracted at prices lower than gas plants (Carbon Brief, 2017). It would be ludicrous to have invested so heavily in subsidies to bring down prices and then not seriously consider whether renewable technology can play an expanding role in the future electricity mix. Britain once turned its back on nuclear power and as a result now relies on France and China to bring the engineering expertise needed to rebuild nuclear capacity in the country. It would be a waste of subsidies to do the same to wind if it can be shown to be a dependable part of a viable lower carbon electricity mix.

Although financing and technology can have a major impact on a wind project, turbines will only produce power when the wind is blowing and this limits their ability to fully decarbonise the electricity system. Opponents of wind will question how, without conventional power stations or storage as backup, Britain can generate reliable electricity when wind turbines are not operating. For example, over the whole of May 2017, wind provided more than 9% of British electricity, but output from wind turbines varied between a high of 7 GW to a low of 0.5 GW. Variability is an undeniable reality of all forms of variable renewable generation, yet supporters

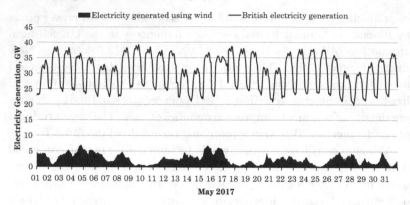

FIGURE 3.4 Electricity generation from Britain's wind turbines over May 2017 were much lower than national electricity demand. The growth potential for wind power alongside other forms of electricity generation is addressed in the Chapter 6 vision for lower carbon electricity (MyGridGB, 2018).

of wind argue that every unit of electricity from a turbine reduces fossil fuel use. The reality is somewhere between what supporters and opponents of wind say. The variable output of wind turbines means that they cannot be the only solution to providing consistent low-carbon electricity in Britain, but there is little doubt among many in the electricity industry that they should be part of a mix of technologies reducing the British impacts on climate change: particularly if costs keep falling (Figure 3.4).

Hydroelectricity

Hydropower generates 16% of global electricity, and some of the biggest power stations in the world are hydroelectric generators such as the Three Gorges Dam on the Yangtze River in China, which at 22.5 GW has more than double the capacity of all of the nuclear power stations in the UK. According to the British Hydropower Association (BHA), there is around 1.7 GW of hydropower capacity installed in the UK, yet in 2017, this provided just 6% of all renewable energy in Britain (DUKES, 2018). Unlike China, the USA or Brazil, there are no large rivers or geographic fall in Britain as needed for large hydropower stations. As a result, hydroelectricity might seem rather insignificant and not worthy of further investigation. That is not an opinion shared by myself because every unit of electricity economically generated in a low-carbon way reduces our fossil fuel usage.

In addition to hydroelectric generators, a key technology in the route to decarbonised electricity is the development of pumped storage hydroelectric facilities. Pumped storage plants use electricity to push water up a hill into a reservoir as a means of storing energy and when that energy is needed, the water is discharged through turbines to generate electricity. Visitors can be shown around pumped storage facilities at Dinorwig in Snowdonia or Cruachan in Western Scotland and marvel at these amazing constructions.

Cruachan is a prime example of supreme engineering effort: the facility rests on the side of Loch Awe with an inlet hiding the opening of two underwater channels. These lead to a vast rectangular cavern which was carved into Ben Cruachan where water is drawn in from the Loch by four large pumps. The cavern, which was constructed as a turbine hall, is a marvel to see with huge pumps with their coloured valves and pipes, being reminiscent of the set from a James Bond film. From the pumps, even longer tunnels turn to rise 396 m towards an upper reservoir. The reservoir itself is contained behind a dam high up on the Ben and which is said to contain fish that have braved the channels and turbine blades to the empty lake above – the evidence of these being a bird of prey who regularly plucks the fish out of this upper lake. When Britain needs electricity, the pumps are switched to turbines and water races from the upper reservoir through into the Loch. The turbines rotate large electric generators, producing over 400 MW of electricity within a few seconds as the upper reservoir is slowly emptied.

Pumped storage facilities like Cruachan have been used to provide a flexible, on demand electricity source since the 1960s. These plants were built in part to provide a black start facility or backup to Britain's nuclear power stations. The fear was that if a nuclear power station had an emergency or fault, then a large percentage of the British electricity supply could be lost at once. Pumped storage was our fast backup that could be brought online to keep the lights on if a nuclear power station failed or to charge up from a nuclear power station if a major load tripped.

In a decarbonised electricity system where the generation cannot directly be controlled, there is a need for assets that can be controlled to match supply to demand. Some hydroelectric generators have a role to play in this flexibility where they can reduce production in response from a signal from the electricity system operator, for example, dams might back off their production on a summer day when solar output is high, and instead run at night when there is a shortfall of solar power. Although different hydroelectric generators have different levels of flexibility, they are an important component in a suite of responsive electricity assets.

Solar power

Solar photovoltaic (PV) panels convert the power from the sun into electricity. Historically, an expensive technology, between 2009 and 2018, the cost of producing useful electricity from solar panels fell by 77% (Bloomberg New Energy Finance, 2018b). Solar panels costs have fallen using mass-production techniques in factories that make hundreds of thousands of panels a year. The panels themselves are now more efficient and produce more power per square metre and work more efficiently in low light. In 2010, the British solar industry was insignificantly small, yet within seven years, nearly 12 GW of solar had been installed on homes, businesses and solar farms. This mirrors strong international growth in the solar industry which has made it an established technology in many countries – as shown for Britain in Figure 3.5.

Solar can be highly effective in Britain when used in the correct application. On homes and businesses, solar panels can directly meet a significant proportion of a consumer's annual electricity supply, particularly where they have a low load or when the solar is combined with battery energy storage – as explored in Chapter 7.

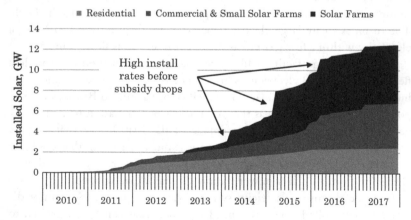

FIGURE 3.5 Deployment of solar photovoltaic in the British electricity system has come from solar farms, residential installs and commercial scale projects. Under subsidies, there was a mass connection of solar assets around March before the Feed-in-Tariff fell, and in 2017, solar generation investment plateaued after the end of subsidies. Falls in solar power prices since 2017 mean that the solar industry is likely to be resurgent from 2019 onwards (Department for Business, Energy & Industrial Strategy, 2018).

Similar to wind turbines, solar panel investments require a large upfront capital with low operating costs. As a result, low costs of finance and bankable agreements to purchase energy from the panels are necessary to keep solar electricity prices low. The solar industry recognises this, and to assure developers, panels regularly come with twenty-five- or thirty-year guarantees on production and high-quality modern inverters (needed to convert electricity from the panels into grid electricity) come with a minimum ten-year warranty. Costs of solar continue to fall with increasing scales of manufacturing; over the last six months of 2018, there was a 20% drop in solar panel prices in some markets. However, although the cost of electricity from solar is low, the technology is only useful when the sun is shining. A passing cloud can reduce the output from a solar farm by over 70% and the production of electricity from photovoltaics during the British summer far exceeds any electricity generation during short winter days. Like wind, solar is variable in its output and like wind that courts controversy around technology acceptance. Solar (and wind) are not the only controversial renewable technology. Bioenergy attracts a fair amount of criticism on environmental and economic grounds.

Bioenergy

Bioenergy is presently used in Britain to reduce the use of fossil fuels in electricity generation. This has been spurned by swapping the fuel used at old power stations from coal to biomass and also the development of new power stations using biofuels such as waste food and sewage. The largest biomass power station is Drax in Yorkshire, which used to be one of the largest coal-fired power stations in the world, but has since converted to one which burns biomass in some of its units. There are valid concerns that overexploitation of bioenergy could affect the viability of this form of electricity generation by putting unsustainable pressure on the natural environment, forests and food supplies. However, increasing the market for bioenergy from less damaging sources such as energy from waste and sewage can have environmental benefits, such as reducing the waste sent to landfill.

Evidence from the Intergovernmental Panel on Climate Change (IPCC) says that bioenergy can be lower carbon than fossil fuels. Accordingly, there is a credible technical and carbon case for high-power biomass plants which are used as flexible backup plants to lower carbon generation. One way of operating bioenergy plants in a highly decarbonised electricity system is to, as much as possible, deploy them to produce energy when fossil fuels would otherwise be used. For example, high winds might mean that wind turbines and other low-carbon power stations could meet all of Britain's electricity demand at the start of a week. However, a drop in wind speeds and an emptying of electrical energy storage plants might mean that there is a shortfall of electricity by the end of the week. In a higher carbon power system, gas plants would be used to meet electricity demand in Britain when wind falls. However, like gas, biofuels can be relatively easy and cheap to store,[2] and in a decarbonised system, bioenergy can be kept in reserve for months at a time and used as a gas substitute when needed. As such bioenergy storage could be viewed as part of an energy storage mix much like (Wilson, et al., 2010) showed is done with coal and gas stores today.

There is also an increasing focus on using bioenergy with other technologies to reduce carbon impacts when they are burned. Carbon capture and storage (CCS) are a collection of technologies which capture and transport waste carbon dioxide from facilities such as power stations and steel manufacturing plants. They then store that carbon dioxide in a way where it will not subsequently be released into the atmosphere to cause a greenhouse effect. It is argued that when CCS is combined with traditional power stations, it makes fossil fuels more environmentally viable by reducing the emission of greenhouse gases. However, this merely perpetuates the burning of finite and unsustainable fuels. Alternatively, CCS can be used on bioenergy power stations meaning that greenhouse gases which are removed from the atmosphere by wood and plants as they grow are permanently prevented by re-entering the atmosphere by the CCS process. This is seen as a means of taking carbon dioxide out of the air. Technical challenges exist to achieve such large-scale bioenergy CCS and socio-environmental issues surrounding biofuels remain. However, should these be overcome or managed, bioenergy CCS has been identified by eminent bodies including the Committee on Climate Change and the IPCC as a key technology for stabilising and reducing human impacts on the climate. Bioenergy with or without CCS can be lower carbon than fossil fuel plants and if used in the right volumes can also be sustainable. For that reason, bioenergy and ideally bioenergy with CCS should form part of the low-carbon toolkit.

A renewable energy mix

There is a finite amount of renewable generation that can be built because there are only so many feasible sites for wind farms, land available for solar farms and roofs for solar panels. Bioenergy requires a feedstock of which there is a restricted amount available without compromising global food supplies, taking more land and water resources for farming or increasing international deforestation. Relying just on wind or solar power alone would leave shortfalls of electricity when the wind is low or overnight or there is little sunshine. Just as with conventional power stations, it is important to install a true mix of renewable electricity generators particularly where these are complementary.

The complementarity of wind and solar is evident in the monthly electricity production of these technologies over a year (Figure 3.6). In Britain, wind is usually strongest in the winter and solar is much stronger in the summer. Solar produced 7.5% of British electricity in June 2018, yet those same panels produced just 0.8% of electricity just six months earlier. Wind also exhibited levels of seasonal and day-to-day variability. To make low-carbon energy work most effectively, it is often found to be necessary to have a balanced electricity mix blending together wind and solar power with a large geographical spread sources to mitigate the variability of individual generators. These can then be backed up through other power sources such as energy storage or gas engines.

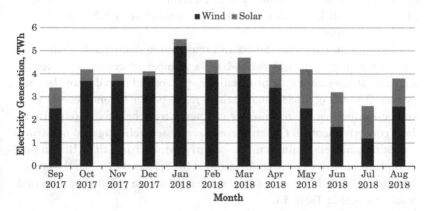

FIGURE 3.6 Production of electricity from British solar and wind from January 2017 to December 2017 show how, on a month-by-month basis, solar and wind capacity could be scaled to provide electricity production which more closely monthly electricity demand. Analysis is undertaken in Chapter 6 of hour-by-hour matching of wind and solar to demand to see how well both can contribute to decarbonisation on a more granular scale [analysis by author using data from (Gridwatch, 2018; Department for Business, Energy and Industrial Strategy, 2019)].

Not all renewable electricity generators have the same variability. Bio-energy and hydroelectricity produce a fairly consistent amount of electricity throughout the year, but there is a restricted growth in both of these technologies due to the availability of biomass and feasible hydro sites in Britain. Tidal power stations are predictable as they mirror the action of the oceans and geothermal plants can extract a constant source of electricity from the heat of the earth and such technologies can only add to the low-carbon toolkit presented here. However, there remains another low-carbon form of electricity generation which produces a relatively stable power output and which is already widely established in the British electricity mix, that being nuclear power.

Nuclear power

Nuclear power stations typically use energy from the decay of materials like uranium and plutonium to generate electricity. The world's first commercial nuclear power station was constructed in Britain in the 1950s and since then the technology has been part of the electricity mix of the country. When running, nuclear plants provide a steady baseload of low-carbon electricity and between January and December 2017 generated around 24% of the British electricity with an output between 5.3 and 8.7 GW (Gridwatch, 2018).

Although they are low-carbon and much more controllable than wind and solar facilities, nuclear plants have a number of critical weaknesses, which mean they must only be installed as part of and not all of an electricity mix. The use of nuclear materials to make power is controversial with some claiming that the technology can never truly be safe. In response to these concerns, Scottish Governments have passed policy against construction of new nuclear facilities in the country. More than half of British nuclear plants are due to be decommissioned by 2025 as they reach the end of their useful life, while a raft of new nuclear plants are planned or proposed as summarised in Table 3.1.

Although more nuclear power is planned in Britain, no nuclear plant has been built in the country since Sizewell B in 1995. The nuclear industry has subsequently faced a skills shortage with insufficient numbers of qualified and experienced engineers to design and construct these plants without international assistance. Britain, one of the pioneers of atomic energy, now has to rely on French and Chinese technology for the next generation of nuclear power plants. These are proving to be expensive and difficult to build and are likely to raise the price of electricity for all consumers.

TABLE 3.1 Existing, planned and proposed nuclear power stations in Britain as of July 2018 (a continually changing table). Nuclear power plants are multi billion pound investments which form key parts of the UK's relationship with countries like France and China for design and construction contracts. This means that there is huge political as well as environmental and monetary capital associated with these projects which ultimately affect their certainty (World Nuclear Association, 2018)

Existing

Nuclear power station	Capacity, MW	Expected closure
Dungeness B	1,040	2028
Hinkley Point B	840	2023
Hunterston B	830	2023
Hartlepool	1,190	2024
Heysham 1	1,160	2024
Heysham 2	1,240	2030
Torness	1,205	2030
Sizewell B	1,195	2035

Planned/under construction/proposed

Nuclear power station	Capacity, MW
Hinkley Point C	3,340
Sizewell C	3,340
Wylfa Newydd	2,760
Oldbury B	2,760
Moorside	4,175
Bradwell (proposed)	2,300

One such power station, at Hinkley Point, will cost twice as much as the London Olympic Games to construct and will produce power at an above market rate: a committee of MPs found that this single plant will increase domestic electricity bills by £10–£15 per year (National Audit Office, 2017). Flawed Government policy in the 1980s which discouraged new nuclear plants have now left Britain exposed to expensive power built by foreign companies. This is potentially good news for other low-carbon electricity generators, as it makes it easier for them to compete. Returns on alternative electricity sources such as wind, solar and bioenergy will be highly profitable if they can sell at the same rate that nuclear power plants of the future will be eligible for.

In addition to cost issues, nuclear power also has a number of technical requirements that need to be considered. As with all power stations, it is important to ensure that there will be sufficient demand to take the electricity that nuclear power plants produce. Of relevance when determining an electricity mix, nuclear power stations should run at a continuous output and only be switched on or off in a slow, measured and controlled process.

The inability to switch nuclear power stations on and off on a regular basis can be restrictive at night or at weekends when electricity demand falls to its lowest levels. The difference in demand between night and day is acute in British electricity; for example, between 16 and 22 February 2017 electricity generation rose above 40 GW at peak, but it fell as low as 26 GW at night. The amount nuclear power stations that can be commissioned in Britain are limited by both the electricity demand and the proportion of that demand which is met by other low-carbon generators. For example, in the morning of 13 February, there was less than 15 GW difference between what nuclear, wind, solar and hydro were producing and what the country was consuming which means that there is only headroom for 15 GW of additional nuclear power stations in this single week (Figure 3.7).

Electricity systems with a high contribution from nuclear power generally have more consistent demand profile as is the case in France where

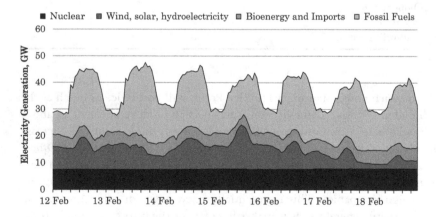

FIGURE 3.7 British electricity mix from 12 to 18 February 2018. A baseload of nuclear power was supplemented by wind and solar. Gas formed the bulk of generation while coal was used during most days. Storage was deployed to meet evening peak demands (MyGridGB, 2018).

nuclear plants provided more than 70% of generated electricity in 2017 (RTE France, 2018) and where night-time demand is more similar to daytime electricity consumption. In the past, a combination of cheap night tariffs, pumped storage plants and streetlights were used in part to ensure there was enough electricity consumption for nuclear power plants to run while Britain slept. These mechanisms are good examples of the tools that planners can use to ensure that the amount of electrical power being generated is matched by consumption on a minute-by-minute basis. In Britain, electrification of heating and transport may well be part of a national strategy to raise demand at night to be able to accommodate that additional nuclear powered electricity. Another mechanism which has a big role in such "balancing" of electricity supplies are the interconnecting cables between the British power grid and continental neighbours.

Interconnectors

The British electricity system is connected to European neighbours using subsea cables known as interconnectors. The cables were installed to allow trading of energy so that, for example, an excess of electricity generation in France can be sold into the British market. In 2017, the year after the Brexit referendum, more than 6% of British electricity was imported from European Union countries through these interconnectors. Under a climate of Euroscepticism and uncertainty about the stability of relations with Europe, some feel that relying on interconnectors is a threat to energy security. Those worries came to the fore during the Brexit negotiations as both parties tried to agree a common framework to allow Britain to continue to import electricity from France, Ireland, Holland and Belgium (Figure 3.8).

Despite the controversies surrounding them, the number of interconnectors to Britain is very likely to increase. In contradiction to some public opinion, regulators feel that interconnectors can actually boost energy security and reduce power bills. By connecting the grid to other countries, interconnectors provide a backup source of electricity should British power stations fail and also allow the import of cheap power from Europe when there is an oversupply of wind or solar power on the continent.

A proposal for an interconnector to Norway will allow Nordic countries to purchase surplus power when there is an oversupply in Britain. Norway will be well placed to buy the electricity because it can store the energy in pumped storage plants. When there is a shortfall of power in Britain or if prices spike, Norway will be able to sell stored electricity at a premium.

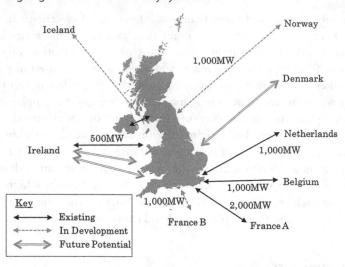

FIGURE 3.8 Status of present, future and potential electricity interconnectors as of December 2018. Interconnectors allow the trading of electricity between different power grids.

Interconnectors add resilience, so if generator fails in one country, it can be backed up by power plants in another electricity grid. This happened in November 2017 when prolonged outages in the French nuclear fleet led to British coal power stations coming back to life to export high-carbon electricity across The Channel (Department for Business, Energy and Industrial Strategy, 2018). Interconnectors are not 100% reliable and their failure can push up prices: on 9 May 2016, some British power stations charged twenty-four times the usual amount to provide backup power when 1.8 GW of power stations and the French interconnector failed at the same time (Crossland, 2016).

From a carbon perspective, interconnectors can have both advantages and disadvantages. Some imported electricity presently comes from European coal power stations; however, in the future, those same interconnectors should allow low-carbon electricity to be traded across Europe as the weather and electricity demand changes across the continent. If wind turbines in Britain are producing more power than needed in the domestic market, excess power can be sold to neighbours rather than those countries using high-carbon power stations. Similarly, if there is a shortfall of renewable electricity generation in Britain, then low-carbon power could be imported from abroad to avoid using fossil fuels at home. This trading works best when there is a simultaneous excess of low-carbon generation in one electricity system and a shortfall in another.

Interconnectors undoubtedly add the flexibility that utilities need to help manage electricity, make power more reliable and potentially reduce bills. As a result, interconnectors are an established component of the British electricity grid, but their contribution to the energy mix is small when compared to fossil fuel power stations.

Thermal electricity generation in a low-carbon energy system

In 2017, gas and coal power stations provided more than 45% of British electricity. Both of these fuels are finite, and by definition, nearly half of the electricity generation in the country is unsustainable. Gas and coal power stations are also by far the most carbon intensive way of making electricity irrespective of the inexcusable effects that fossil fuel extraction, transportation and combustion is having on the environment and geopolitics.

To meet UK net-zero carbon targets, the Committee on Climate Change estimates the average grid intensity of electricity generated in 2030 should meet an interim goal of between 50 and 100 gCO_2eq./kWh (Committee on Climate Change, 2018). This target should be contrasted with information from the IPCC which shows that without any form of CCS, coal power stations produce between 740 and 910 gCO_2eq./kWh and gas power stations produce it in the range 410–650 gCO_2eq./kWh. Clearly, if any country is to meet climate goals, it cannot rely entirely on coal or gas as the carbon emissions are too high for the world to sustain.

The UK government has committed to phasing out coal power by 2025 (Cockburn, 2017). In parallel with a coal phase out, gas use must now fall to achieve the required 100 gCO_2eq./kWh carbon intensity by 2030. For that to occur, the required rate of decline of coal and gas in the electricity grid can be calculated as shown in Figure 3.9. By 2030, just 7.8% of British electricity generation a year can come from gas if a interim goal of 50 gCO_2eq./kWh is to be achieved. From a carbon perspective, gas might be permitted to provide 50% of our electricity for a few days a year as long as there are sufficient days where it is providing much less to keep the annual average contribution below 7.8% of the total electricity generated. In short, gas or coal can be used to generate electricity, but only in much small amounts if medium-term carbon goals are to be met. The decarbonisation seen in recent years needs to continue if carbon goals are to be met.

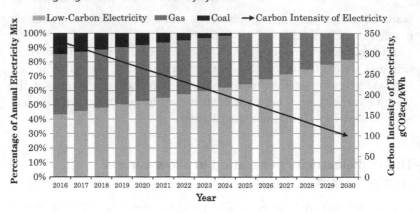

FIGURE 3.9 Rate of decarbonisation required for British electricity to have a carbon intensity below 100 gCO$_2$eq./kWh by 2030. This simplifies how coal and gas need to be less prevalent and replaced by low-carbon electricity sources every year until and beyond 2030 [analysis by author, similar to requirements for decarbonisation in (Committee on Climate Change, 2018)].

Is this practically achievable? Many are presently promoting natural gas, the lowest carbon fossil fuel, as the ideal fuel to transition from a high-carbon to a low-carbon electricity supply. In decarbonised Britain, a worst case event would be a cold, dark and still winter day when the wind is not blowing and the sun is low in the sky. As a result of this poor weather, the energy produced by solar and wind plants might be insufficient to meet our demands, and such shortfalls in electricity production could easily be compounded by unscheduled maintenance on nuclear plants. Regardless of the weather, Britain needs sufficient power to run hospitals, schools and businesses, throughout the year. To meet any shortfalls when renewables are not generating, some form of electricity generation needs to be switched on, and there is an argument that this role could partially be fulfilled by gas power stations. Under such a scenario, gas power stations might not be used all of the time. For days when it is windy or sunny, then it would be prudent to have wind turbines and solar panels installed in order to capture the good weather. Similarly, other low-carbon power stations have a role in generating power and reducing the rate at which fossil fuels are consumed. The argument for sustainable electricity is often made on making an energy system which is not ruinous to our planet. To rule out building wind, solar, tidal and hydroelectric assets for other times of the year when it is windy, sunny, etc. means missing an opportunity to switch gas stations off and, as a consequence, saving fossil fuels for those cold, dark and still winter nights. The good news for the climate is that cold winter nights do not happen all year around, and British weather suits the use of wind and solar plants for a large part of the year.

With our finances, it is responsible to save money for when you really need it rather than to spend it all in one go. Fossil fuels are a finite resource with a million useful purposes besides energy, and, much like our money, it is a waste to burn them rapidly when low-carbon alternatives exist. Unfortunately, fossil fuels are burned in Britain without regard for the sustainability of the planet or the long-term finances of people.

Gas turbines have technical and economic advantages which make them viable as a last resort backup to renewables. It is presently much easier and cheaper to store gas in large volumes than it is to store electricity in the equivalent number of batteries or pumped storage plants. This means that it is easier to keep a strategic reserve of fossil fuels ready to be burned in power stations. Second, the economics of gas power stations mean that compared to renewable generators they have a relatively low installation cost and a more expensive operating cost. As such, the commercials that underpin gas power are more suited to a plant that is run occasionally rather than all of the time.

The key to designing a lower carbon energy mix is having enough variable renewables and storage that you rarely need dispatchable power stations. However, using thermal plants such as diesel, gas and coal as dispatchable power sources rather than as baseload fundamentally changes the commercial models that these power stations run on. Today, fossil fuel power stations make money by almost continually burning fuel to generate electricity, yet in the future, their role will be the provision of power only as backup to low-carbon power stations. This is a stark change, but one which might be necessary to achieve decarbonisation. This fundamental change to the commercial models for thermal power plants is something which is, no doubt, of great interest to the whole energy industry. It presents unique engineering and economic challenges in producing commercially viable thermal engines which will be operated for tens rather than thousands of hours a year. It also presents interesting challenges around how and when the fuel for the thermal plants is generated/manufactured/refined and how that fuel is stored.

Fuel storage has played an important role in the history of British electricity by allowing fossil fuels to be mined and stored as reserves before being used for power generation (Wilson et al., 2010). In the future, our gas power stations might need large reserves of fuel which are ready to be burned when wind or solar output is low. Unfortunately, Britain might not be preparing for this scenario because in 2017, the Rough gas storage facility (which accounted for 70% of British gas storage) was scheduled to be closed. This closure is reported by some to be an economic necessity (McKinsey Energy Insights, 2017), but these storage facilities might be brought back into use if gas storage becomes valuable again.

These technical and economic challenges do not just apply to gas power stations as they are relevant to most forms of thermal power where a fuel is burned to make electricity. Indeed, they may present opportunities for other forms of dispatchable power. This includes bioenergy plants where it is easy to store the fuel and to produce feedstock throughout the year. Although there is only a finite amount of bioenergy that can sustainably and responsibly be produced every year, could the use of thermal plant be reduced so much that only sustainable biofuels are used in them? Britain and other countries might find that the volume of energy required from thermal plants might be reduced so much that biomass or biogas plants become viable wholescale replacements of gas or coal. Image if biogas, for example, could be created during the summer months to be burned as a lower carbon alternative to gas in the winter.

Demand side tools

Demand side refers to mechanisms to shape or reduce consumption of electricity to permit a lower carbon or lower cost power system. Intuitively, the smaller the demand for electricity, the less electricity needs to be produced. This results in carbon savings when less fossil fuels are used in power stations or fewer electricity generators are built. However, the interdependence of electricity demand and carbon gets increasingly complex as the power system decarbonises, meaning that demand response does not always directly correlate with greenhouse gas reductions.

Reducing electricity demand might come through energy efficiency improvements of appliances and machines that are in use every day. More efficient equipment uses less electricity, but the resulting carbon savings might be offset by higher greenhouse gas production (or indeed pollution) in manufacture. Indeed, higher efficiency appliances might use unsustainable materials or design which place increased stress on global mineral resources. Demand reduction often has consequent social and environmental effects. One of the drivers for demand reduction through efficiency is higher electricity prices which make energy saving more valuable. These higher prices might help mitigate the effects of climate change though reducing demand, however they can also negatively and disproportionately impact those with low incomes who are already in or approaching fuel poverty.

Demand side tools need to be smart to have the most positive benefits. In a power system with large amounts of solar generation, such as one in the middle of a desert, there will be an abundance of low-carbon electricity during the day while at night, electricity might be provided by a combination of batteries and diesel generators. Reducing electricity demand in the daytime has a number of effects; on the one side, it means there is more of a surplus of solar power which can be used to charge batteries, and these fully charged batteries can subsequently be used to reduce use of diesel generation at night; however, on the other hand, it might also be the result of consumers shifting their power demand to the evening when a generator needs to be used. In such desert power systems, it is most desirable to reduce electricity demand at night-time as this is when it is most likely that a diesel generator will be used. Using less energy at night also means a small battery might be needed, and a small battery means fewer manufacturing derived carbon emissions. In all power grids, shaping demand to match low-carbon generation is a key method for helping reduce greenhouse impacts of electricity.

In a world with a large number of electric vehicles, shaping demand could become increasingly critical to protect networks and reduce carbon emissions. If all electric vehicles are allowed to charge at the same time, it would place an extraordinary demand for electricity on networks and generators. As a result, a large number of power stations would need to be used to meet demand and larger cables installed to get that high power to vehicles. Smoothing out the charging of vehicles to reduce the peak demands on the power system and matching charging more closely to the generation of low-carbon electricity will evidently be vital for decarbonising transport and power.

These examples illustrate just a few reasons why demand reduction is complex but also potentially very valuable. As Britain decarbonises, demand side tools should certainly be in the low-carbon toolkit. However, these must be designed just as carefully as any other intervention to replace fossil fuels.

Designing decarbonised electricity systems

At times of major policy announcements for decarbonisation, it is important to be able to judge whether these can credibly help Britain achieve carbon targets. Can present policies lead to a low-carbon electricity future be built? How effective is a low-carbon toolkit? How sustainable can electricity become?

Over the next few pages, an evaluation is performed of some of the measures in the low-carbon toolkit to see if any of them could cause significant reductions in greenhouse gas emissions. To simulate alternative electricity supply scenarios, a simple model of the electricity system is built using historical data from the British system. This data shows what electricity demand was for each hour of the day, which power stations are available to meet that demand and how much wind and solar resource there was. To simulate how to meet that demand, electricity generators are added one by one in a hierarchy until the total output of these is enough to meet the demand. The *merit order* of electricity generators is nuclear (as it must always run at a consistent output) followed by wind, solar, hydro and then energy storage. Electricity sources such as biomass and imports are then dispatched before gas and coal.

To determine how much renewable generation is available, I look at the historical data and scale it according to a future mix. For example, if there is twice as much solar in a simulated electricity mix, then the amount of solar in the historical data is doubled. If the total amount of low-carbon generation potential exceeds national demand, then the model switches off generators so that supply matches demand. If there is not enough low-carbon generation, then gas plants and energy storage are brought online to match demand (Figure 3.10).

The model is about demonstrating whether there is practically enough low-carbon electricity generation to meet Britain's electricity demand, rather than presenting all of the detail of how that might be achieved. The model does not attempt to be sufficient for power systems engineers working out all of the details of how to power the country on a minute-by-minute basis, but it is sufficient to determine the viability and impact of different scenarios on decarbonisation. Some of the assumptions used to make this model are provided in the appendix.

British electricity in October 2017

In October 2017, around 50% of British electricity came from low-carbon sources including nuclear, hydro, low-carbon imports, wind and solar. Britain's nuclear power stations provided a steady baseload of low-carbon energy despite some of these power plants being shut down for maintenance. Wind was variable, although it rarely dipped below 5 GW of total generation while solar output at midday peaked above 7 GW at the start of

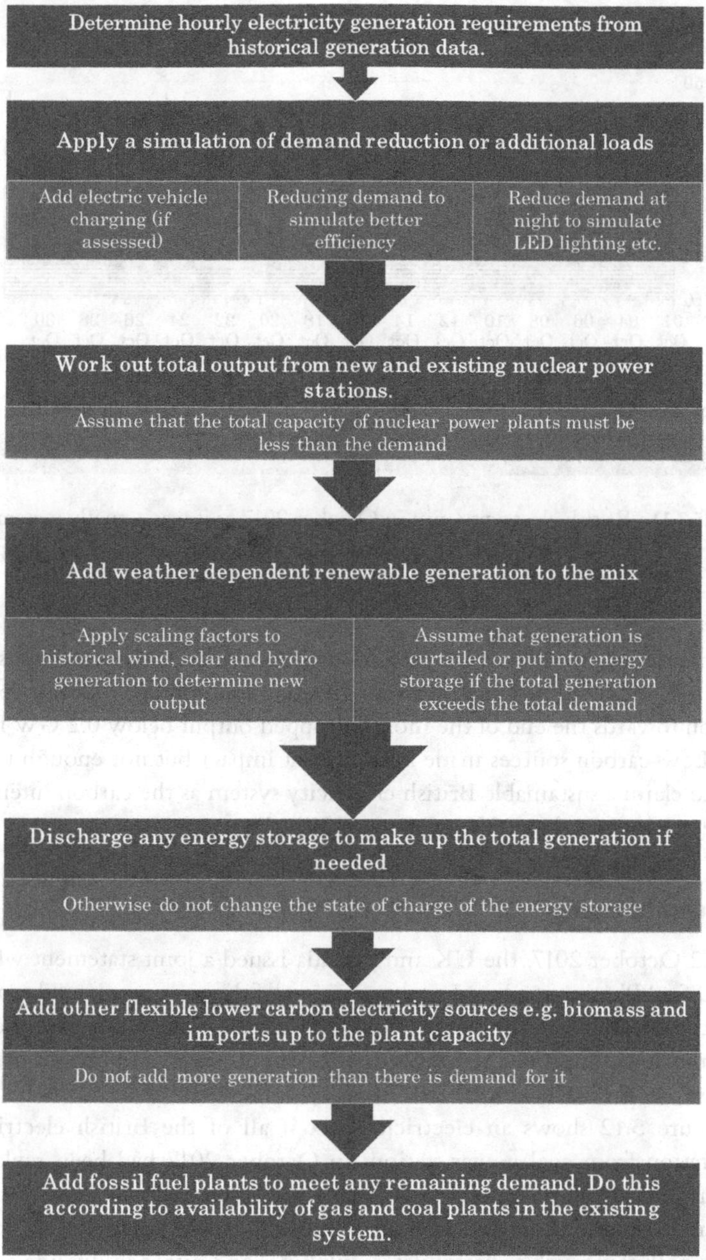

FIGURE 3.10 Overview of the merit order plant dispatch model used to simulate an electricity system with a different electricity generation mix.

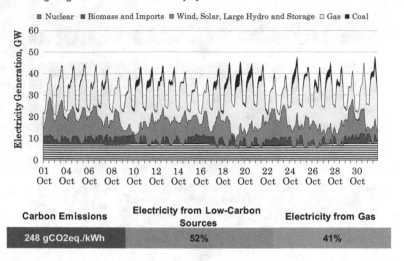

FIGURE 3.11 British electricity mix in October 2017 as it was actually generated. Low-carbon electricity sources provided more than 50% of supply, but carbon emissions were still very high (MyGridGB, 2018).

the month but dipped below 2 GW towards month end. Biomass plants averaged 0.75 GW of production although a day-long outage of Drax Power Station towards the end of the month dropped output below 0.2 GW for a day. Low-carbon sources made a significant impact but not enough to be able to claim a sustainable British electricity system as the carbon intensity of electricity was above 250 gCO$_2$eq./kWh (Figure 3.11).

The end of coal

On 12 October 2017, the UK and Canada issued a joint statement which called for "Phasing unabated coal power out of the energy mix and replacing it with cleaner technologies will significantly reduce our greenhouse gas emissions, improve the health of our communities, and benefit generations to come" (Cockburn, 2017).

Figure 3.12 shows an electricity mix if all of the British electricity generation from coal power stations in October 2017 had been replaced by gas. On the surface, this looks like a really progressive move for the Britain because carbon emissions are reduced to below 200 gCO$_2$eq./kWh while pollutants such as particulates, sulphides and nitrogen dioxide emissions from coal plants are not released into the environment. However, the amount of fossil fuel electricity remains above 40%, British electricity

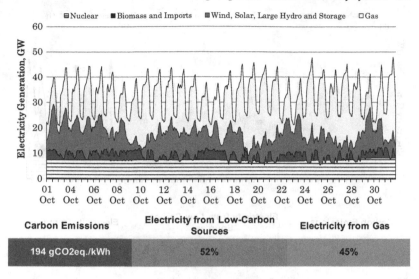

FIGURE 3.12 British electricity mix in October 2017 with coal replaced by gas. Replacing coal with gas reduces the carbon intensity of electricity but emissions are still unsustainable. Further, this leaves Britain even more reliant on an unsustainable fossil fuel in natural gas.

remains unsustainable and carbon targets are missed. The policy is an improvement but not a significant enough shift to cause anything like what is required. As Green MP Caroline Lucas said in 2015,

> This switch from coal to gas is like trying to go dry by switching from vodka to super-strength cider – it entirely fails to seriously address the real challenge at hand. Investing in renewables and energy conservation would be far more effective economically, environmentally and in terms of energy security. We must begin weaning ourselves off gas as quickly as possible.
>
> *(Mason, 2018)*

Growing low-carbon renewable energy sources

Another policy being enacted in Britain is the installation in renewable power stations. Over October 2017, wind, solar and hydropower plants actually provided over a fifth of British electricity. If this generation is scaled up, it is possible to see what growth in wind and solar could have brought, considering

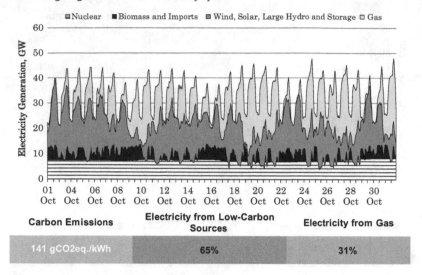

FIGURE 3.13 British electricity mix with growth of wind and solar and no coal. Investment in wind and solar, although intermittent, does further reduce carbon emissions. However, the variability of these still means gas is frequently used to make electricity making carbon emissions unsustainable.

that it is not always windy and not always sunny. Figure 3.13 shows the impact on the energy mix by scaling wind, solar and hydropower generation by one and a half times and keeping coal plants offline. This change is enough to offset most of the gas used for electricity for several days, for example, during the first week of data shown. However, gas remains a major part of the system, providing over 30% of electricity over the whole month.

The introduction of more wind, solar and hydroelectricity would have increased the amount of low-carbon sources and reduced greenhouse gas emissions. However, carbon targets are still missed, and gas is the major electricity provider of electricity on days like 9 and 19 October where there is little wind and sun.

Building new nuclear power stations

Nuclear power can be assessed in a similar way to renewable energy. If 70% of the proposed nuclear plants listed in Table 3.1 are actually built, then there would be a total of 11.5 GW of nuclear power stations to replace the existing nuclear fleet. Such a scenario is shown in Figure 3.14 with the

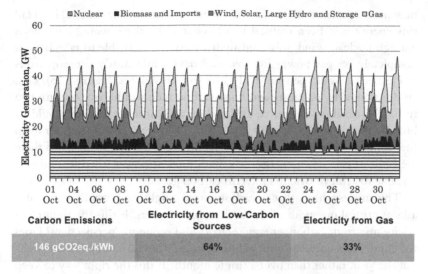

FIGURE 3.14   British electricity mix with additional nuclear power stations and no coal. Nuclear power growth of this magnitude has a similar effect to the wind and solar scenario previously assessed in reducing carbon emissions. However, as with wind and solar, gas is still frequently used and carbon emissions are unsustainable. Nuclear power is simulated using the same outages/power fluctuations as actually occurred in October 2017.

baseload of new nuclear power stations. This is found to achieve similar carbon savings as adding wind, solar and hydro capacity to the grid as the renewables which were assessed in the previous example.

In October 2017, there was actually enough demand for even more nuclear power than the 11.5 GW simulated. If all of Britain's proposed nuclear plants were commissioned, then British electricity would have achieved carbon emissions below a value of 100 gCO₂eq./kWh target in October 2017. However, to value an electricity mix, lower cost electricity and negative public opinion towards nuclear power, it might be appropriate to blend new nuclear with new renewable generation and measure the impact that has on carbon emissions.

Creating an energy mix

Stable, affordable and reliable electricity systems do not correlate with policy which favours any single type of electricity generation. An alternative scenario of creating an energy mix is much more credible such as one with new nuclear, a removal of coal and investment in variable renewables. Had this energy mix been realised in October 2017, there would have been enough nuclear, wind, solar and hydroelectricity available to take Britain completely off gas power for two full days while nearly three quarters of electricity would be from truly low-carbon sources (Figure 3.15). Gas, an unsustainable, polluting and high-carbon fuel, would have transitioned from a baseload power source to a backup and carbon emissions would be brought to a value of 100 gCO$_2$eq./kWh. The inconvenient truth for fossil fuel companies is that despite the variability of wind and solar, when these are built as part of an energy mix with nuclear power, then it might be possible to achieve Britain's interim carbon goals.

This analysis hints that decarbonisation might be possible, but it also asks as critical questions of electricity policymakers. Is this the right electricity mix from carbon, energy security and economic perspectives? Does the energy mix correlate with public opinion? Can this mix work throughout the year, rather than over a single month? Is this the right way to keep bills low? In Chapter 6, an alternative energy mix is presented which can achieve carbon emissions targets over an entire year.

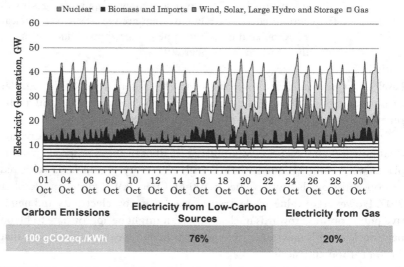

FIGURE 3.15 British electricity mix with growth of wind, solar and nuclear and no coal. Gas is used much less frequently and carbon emissions are the lowest due to the volume of low-carbon energy. This simulation only looks at one month of the year. A key question remains as to what such diversification of electricity generation can do for annual carbon emissions and gas use.

Conclusions

This chapter has discussed how traditional tools are being utilised to construct a more sustainable electricity system. Through evaluation of a single month, October 2017, growth in low-carbon energy was shown to achieve a more sustainable electricity supply for Britain, with natural gas used when it was needed rather than as a mainstay of energy generation. However, the approach did not evaluate whether this was the right or most likely way in which decarbonisation will occur, and it did not examine how well the electricity grid might cope. The next chapter will look into one technology which might help make a low-carbon future work, electrical energy storage, and what might bring about the large volumes of that technology. Welcome to the store-age!

Notes

1 New Zealand can go for months generating more than 55% of its electricity from hydropower. The remainder of the country's electricity comes from geothermal, gas, coal and wind.
2 If the bioenergy is stored as fuel in tanks or bioenergy dumps.

References

Annan, K., 2015. *We must Challenge Climate-Change Sceptics Who Deny the Facts* [Interview] [03 05 2015].

Bloomberg New Energy Finance, 2018a. *France Onshore Wind Auction Sees Price Drop.* [Online] Available at: https://about.bnef.com/blog/france-onshore-wind-auction-sees-price-drop/ [Accessed 31 01 2019].

Bloomberg New Energy Finance, 2018b. *Tumbling Costs for Wind, Solar, Batteries Are Squeezing Fossil Fuels.* [Online] Available at: https://about.bnef.com/blog/tumbling-costs-wind-solar-batteries-squeezing-fossil-fuels/ [Accessed 25 08 2018].

Carbon Brief, 2017. *Analysis: UK Auction Reveals Offshore Wind Cheaper than New Gas.* [Online] Available at: www.carbonbrief.org/analysis-uk-auction-offshore-wind-cheaper-than-new-gas [Accessed 25 08 2018].

Cockburn, H., 2017. *UK Vows to Close All Coal Power Plants by 2025.* [Online] Available at: www.independent.co.uk/environment/coal-power-plants-uk-close-2025-renewable-energy-amber-rudd-nuclear-gas-policy-a7997241.html [Accessed 18 08 2018].

Committee on Climate Change, 2018. *Reducing UK Emissions: 2018 Progress Report to Parliament,* London: Committee on Climate Change.

Crossland, A., 2016. *Back-Up or Pack-Up: How the Storage Revolution can Help Keep the Lights On.* [Online] Available at: https://blueandgreentomorrow.com/features/solarcentury-discusses-renewable-energy-storage/ [Accessed 17 06 2018].

Department for Business, Energy & Industrial Strategy, 2018. *Solar Photovoltaics Deployment*. [Online] Available at: www.gov.uk/government/statistics/solar-photovoltaics-deployment [Accessed 11 08 2018].

Department for Business, Energy and Industrial Strategy, 2018. *Digest of UK Energy Statistics (DUKES): Electricity*, London: s.n.

Department for Business, Energy and Industrial Strategy, 2019. *National Statistics: Energy Trends: Renewables*. [Online] Available at: www.gov.uk/government/statistics/energy-trends-section-6-renewables [Accessed 30 01 2019].

DUKES, 2018. *Chapter 6: Renewable Sources of Electricity*, London: Department for Business, Energy & Industrial Strategy.

Electricity Authority, 2018. *EMI*. [Online] Available at: www.emi.ea.govt.nz/Wholesale/Dashboards [Accessed 18 01 2018].

Gridwatch, 2018. *G.B. National Grid Status*. [Online] Available at: www.gridwatch.templar.co.uk/ [Accessed 18 08 2018].

Mason, R., 2018. *UK to Close All Coal Power Plants in Switch to Gas and Nuclear*. [Online] Available at: https://ourworld.unu.edu/en/uk-to-close-all-coal-power-plants-in-switch-to-gas-and-nuclear [Accessed 25 08 2018].

McKinsey Energy Insights, 2017. *Business as Usual for UK Gas Despite Rough?* [Online] Available at: www.mckinseyenergyinsights.com/insights/business-as-usual-for-uk-gas-despite-rough/ [Accessed 13 10 2018].

MyGridGB, 2018. *MyGridGB*. [Online] Available at: www.mygridgb.co.uk [Accessed 17 06 2018].

National Audit Office, 2017. *Hinkley Point C*, House of Commons: National Audit Office.

Reuters, 2018. *France's EDF Halts Four Nuclear Reactors Due to Heatwave*. [Online] Available at: www.reuters.com/article/us-france-nuclearpower-weather/frances-edf-halts-four-nuclear-reactors-due-to-heatwave-idUSKBN1KP0ES [Accessed 13 10 2018].

RTE France, 2018. *2017 Annual Electricity Report*, France: RTE.

Wilson, I., McGregor, P. & Hall, P., 2010. Energy Storage in the UK Electrical Network: Estimation of the Scale and Review of Technology Options. *Energy Policy*, 38(8), pp. 4099–4106.

World Nuclear Association, 2018. *Nuclear Power in the United Kingdom*. [Online] Available at: www.world-nuclear.org/information-library/country-profiles/countries-t-z/united-kingdom.aspx [Accessed 19 08 2018].

4

WELCOME TO THE STORE-AGE

The energy storage mix and its role in flexible electricity

When talking about democratising banking, the CEO of PayPal, Dan Schulman, said that "*Personal and mobile computing, long-distance communications, energy storage, and air travel are just a few of the things that have been democratized by technology, creating new possibilities for billions of people*" (Schulman, 2015). Irrespective of my views on banking, that sentiment could certainly be true in the power sector where battery storage is helping to make island nations, homes and communities less reliant on foreign or fossil fuel-derived energy sources. The benefits of energy storage do not just exist within the realm of democratising energy – it is a technology family which is allowing engineers to bring about a reliable low-carbon transition in power systems around the world.

Power system impacts: can the grid cope?

Low-carbon generators like wind and solar fundamentally change the way that power grids work. The British electricity grid was mostly constructed in the 1960s and 1970s before the present era of decarbonisation and was engineered imagining power flowing in one direction: from large power stations down to homes, businesses, schools and hospitals. One route to decarbonised electricity, using a low-carbon toolkit, is to distribute thousands of small power stations in the form of wind, solar, hydro and marine technologies all over the country owned by both utilities and consumers. Those generators will also be distributed through the grid and close to homes and businesses. This pushes electric current in new directions through the grid

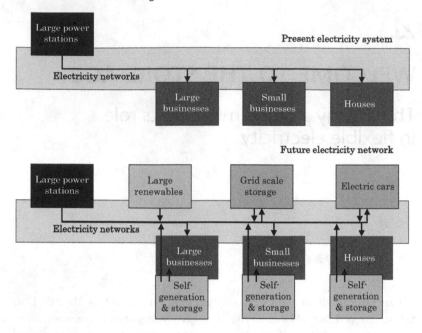

FIGURE 4.1 The conventional and future power grids will be fundamentally different. The present electricity has a small number of controllable assets, while proponents of the future electricity system argue that it has much more devices and so much more flexibility for operators to balance supply and demand, reduce costs and maintain power quality.

and in ways that the power networks were not designed to cope with. For example, solar panels on a home might mean that a house exports electricity into the power system for some of the time, rather than always taking from the grid (Figure 4.1).

There is a huge amount of complexity in transitioning utilities from managing a few hundred power stations to also working with millions of smaller, distributed plants across the grid. Large generators produce electrical power in a consistent, controllable and predictable way which can be completely different to the way wind and solar plants operate. The variability of renewable power in a grid is forcing utilities to find new ways of ensuring that there is always enough power to meet demand.

Designing an electricity grid from scratch to work in a decarbonised way might be relatively easy for today's engineers as is proven by decarbonising of grids around the world; however, the issues that utilities are facing in decarbonising large economies are compounded by the need to transition large and fully working electricity system over to the new way of operation without being allowed to interrupt supply or reduce power quality. The challenge of decarbonisation is not just building affordable low-carbon generation; it is retrofitting that to an operational electricity grid which costs billions of pounds to build and operate. In addition to new generation types coming online, electrification of heat and transport could increase electricity demand beyond levels that the grid was designed to cope with. Power networks were designed when the highest loads were water heaters yet a fast-charging electric car can consume more than ten times the power of a kettle. Many engineers are asking whether the electricity grid can facilitate decarbonisation without billions of pounds of investment.

Increases in peak demand on networks might occur at the same time that the production of electricity gets more variable. So what happens in a decarbonised electricity system if there is a sudden shortfall of power when a wind turbine trips or demand suddenly increases? In the 1960s, British engineers faced a similar challenge in needing to find a way to produce large amounts of power in a very short amount of time. This was due to events such as a mass switching on of British kettles during the adverts between TV shows. The power stations used to make electricity at the time were great at producing power in a stable and consistent manner, but their power output could not be ramped up quickly enough or in a commercially viable way to meet very sudden increases in demand. To address this, engineers constructed large pumped hydro energy storage plants in the mountains of Scotland and Wales. These plants still operate today, as charted in Figure 4.2, which shows how storage was used in Britain in October 2017 to meet morning and evening peaks.

Balancing electricity is not just done using storage. Conventional power systems have a property which is analogous to *inertia*, which allows them to automatically respond to the instantaneous changes in electricity demand which happen every time appliances are turned on or off. This inertia has been critical to keeping the supply and consumption of power in balance. To provide some of that that inertia in a decarbonised world, electrical energy storage is now being deployed which can quickly be charged and discharged to meet changes in supply and demand. If there is insufficient wind or solar power to meet demand, even if for a fraction of a second, then utilities can call on new storage plants to meet the shortfall. Historically, storage has been expensive and difficult to justify on economic grounds. Thankfully, in addition to a changing electricity generation landscape in the low-carbon toolkit, there is fundamental change in energy storage which might allow that to enable decarbonising of electricity.

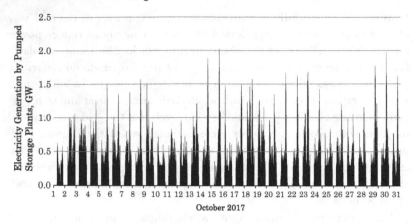

FIGURE 4.2 Contribution of electrical energy storage to the British electricity system in October 2017. Storage is used in the day to generate electricity, particularly during the evening peaks when demand spikes. This is because storage can quickly and economically be deployed for short periods of time, unlike power stations which ideally need to run for many hours at a time to be profitable (Gridwatch, 2018).

What is energy storage?

Electrical energy storage is a group of technologies which are used to contain electricity before it needs to be used. The most iconic and well-known form of energy storage is the battery, but the discipline of energy storage covers a multitude of technology options including various battery chemistries, flywheels, supercapacitors, water reservoirs and hydrogen. Energy storage itself is not just limited to electricity; thermal batteries such as hot water tanks convert electricity to heat and allow that heat to be used when needed.

Storage is useful in a sustainable system as it allows electricity to be produced, stored and used at completely different times (Figure 4.3). For example, electricity produced on a windy day can be stored in a large battery to be used on a few days later when wind turbines are not producing sufficient power or to meet peak power demands in the evening. This "energy shifting" role is one of the most cited applications for storage yet utilities from the USA, China, Europe and Australia have found many other uses for the technology which gives it a wide-ranging and flexible role in the energy system. For electricity system operators, storage can provide *backup services*. If a major electricity station fails, batteries can be used to

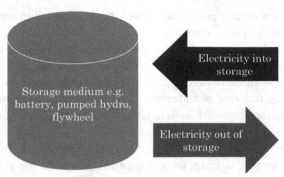

FIGURE 4.3 Simple image of an electrical energy storage device showing how electricity is charged into a storage medium which can then be discharged, when required, to provide electricity again.

provide emergency power and keep the lights on until another electricity generator is turned on. Storage is particularly good at this role as it can be called upon in fractions of a second, whereas gas power stations can take several minutes to come online and a coal power stations can take hours. Storage can be used to reduce losses and increase efficiency when electricity networks are at peak load. Here, a battery might be discharged to meet some of the electricity demand of a customer and reduce demand on the grid. This *peak shaving* application has already manifested itself commercially by some major electricity consumers who now use batteries to avoid buying electricity from the network when prices are highest. For network operators, batteries have been successfully used to add capacity to major substations and reduce peak loads. For example, if a substation can provide 10 MW to a customer, but the customer requires 12 MW for a short time, then a 2-MW battery can be installed to make up the difference and provide power in parallel with the substation. This can be highly profitable where it defers upgrades of expensive electricity network infrastructure.

Community energy groups can use batteries to provide an *alternative to grid electricity*. In these applications, storage can also be coupled with renewable generation to replace or reduce the use of fossil fuels. This is often viable on islands that depend entirely on diesel generators for electricity but are now switching to solar, batteries and generators as an alternative energy mix. Batteries can even be coupled directly to a diesel generator and charged/discharged to maximise fuel efficiency and reduce operating and maintenance costs. The latter can have a payback period of less than four years in some cases. In the UK, storage can be useful for remote villages where there are regular grid outages and who want a backup supply as well as to generate their own electricity.

For customers connected to the grid, storage can be used to buy and store electricity when it is cheap. This stored energy can then be used by the customer as an alternative to grid electricity when prices are high. This *trading/arbitrage* function of batteries has been heavily investigated in markets with variable, low-carbon generators because the influx of renewables tends to cause prices to be much more volatile.

Storage is not just game-changing for a decarbonised electricity system by allowing energy shifting, i.e. storing solar power for use at night. Energy storage brings new tools to make the electricity system more flexible, reliable, stable and potentially affordable.

Electrical energy storage is most valuable when a single installation is used to provide more than one service to customers and utilities. Early adopters of storage in the UK have used a single battery for shifting solar from day to night, reducing peak electricity consumption and for providing backup services to National Grid. The fact that storage plants can be used for many services makes them valuable assets, and as a result academics and industry specialists have found that they can even reduce the costs of running power systems.

Storage is as old as the grid itself. A paper published by Dr Grant Wilson and others (Wilson et al., 2010) showed how other forms of storage have provided key buffers in the electricity system to decouple when energy supplies are harvested from when they are used. Energy stores in the form of gas towers in our cities and coal heaps next to power stations mean that fossil fuels do not have to be extracted at exactly the same rate that they are consumed. As recently as 2009, these stores contained enough energy to keep power stations running for months without fuel resupply. During the miners strikes of 1984–1985, it was these coals stores that kept the electricity system running despite a huge fall in production.

Such buffering is used in everything from food to construction. A tin of beans actually has a lot in common with the way that the electricity system could function in the future. Each of the ingredients in the tin and all of the minerals in the packaging were harvested, processed and stored months before they entered your kitchen. Beans do not need to be produced at the moment; they are used because food suppliers are experts at building economically viable buffers between production and consumption. Food supply is designed to avert starvation by producing crops during growing seasons, harvesting them when they are ready and passing onto consumers when there is demand. Decarbonised electricity might work in a very similar way. To harvest low-carbon electricity as is done with fossil fuels or beans, it is necessary to generate, store and then process low-carbon electricity in a flexible manner, i.e. to decouple production from consumption. This chapter looks at the technologies and cultural change that might introduce that more of the flexibility to work with low-carbon generators. In it, we build up an "electricity storage mix" for Britain starting from pumped hydro through to batteries, electric cars and domestic hot water.

Starting the energy storage mix: pumped storage

One of the oldest forms of storing electricity is pumped storage. Britain has just a handful of such plants in North Wales and Scotland, and most were commissioned in the 1960s and 1970s. As the proliferation of re-newable energy has increased in the UK, there have been consequent calls for a renaissance in the pumped storage industry to bring more storage online.

Correctly installed pumped storage plants have few negative environ-mental consequences and can even be used to rehabilitate damaged en-vironments. They have a long operational life, help to diversify rural economies and they do not require the use of rare or exotic minerals in their production like some batteries. As previously noted, pumped storage plants at Dinorwig in Wales and Cruachan in Scotland are prime examples of remarkable engineering achievements which have helped the local tour-ism industry and provide jobs to civil, electrical and mechanical engineers to keep the plants running.

Much of the original assessment of the viability of pumped hydro plants was completed in the middle of the 20th century, which was a fundamen-tally different time for British electricity. Many sites were found to be uneconomic and increasing the cost of electricity. The economics of to-day's decarbonising power system with an expanded electricity grid should mean that pumped hydro feasibility will be reassessed. The commercial and technical requirements to manage an electricity system with high volumes of variable renewable generators might even make pumped stor-age plants which were previously considered unviable entirely appropriate in the 21st century.

Other than economics, one of the main difficulties surrounding pumped storage, particularly in the UK, is finding suitable geography to build them. Plants have quite particular geological and topographic requirements, such as that required to provide an upper reservoir that will hold millions of gallons of water at sufficient height above a lower reservoir. Sites also need grid connections which are able to pull and export large amounts of power up to twenty-four hours a day which can mean laying cables across some of the most sensitive places in the country. Modern tunnelling or cable burying techniques might make the cabling unobtrusive, but this can be expensive, however the fundamental requirements of geography mean that few sites for pumped storage have been identified (Figure 4.4).

In 2017, a new pumped storage facility was announced for Britain. To be built using old slate quarries in Snowdonia, North Wales and costing £120 million, it would be about a quarter of the price per unit of domestic battery storage systems in 2017 but have an operating life of at least 120 years. This project shows that with some creating thinking, it should be possible to find new sites for pumped storage. On a visit to the one pumped storage facility, I was even shown plans to expand the plant that

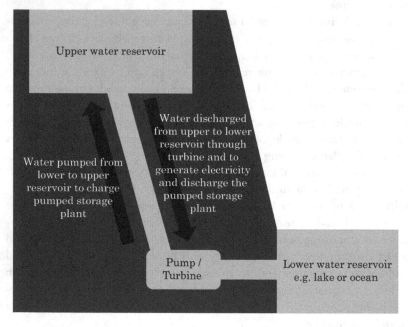

FIGURE 4.4 A schematic representation of a pumped storage plant with an upper and lower reservoir separated by a turbine/generator hall carved into a mountain. The energy capacity of pumped storage plants depends on the volume of water that can be stored in the upper reservoir.

have been in place since it was designed but have yet to reach economic viability. Table 4.1 lists some of the proposed pumped storage projects in Britain, which would provide nearly over three times the pumped storage reserve that the country presently has (Figure 4.5). However, if all of these pumped storage plants are built, the stored energy would only meet British electricity demand for less than four hours. Clearly, more flexibility needs to be found for a decarbonised electricity system as part of our energy storage mix.

TABLE 4.1 Operational, proposed and planned pumped energy storage plants in Britain (DNVGL, 2016; MacKay, 2007; The Engineer, 2017)

Plant	Power, MW	Energy, MWh	Status
Snowdonia pumped hydro	100	700	Proposed
Ffestiniog	360	1,300	Operational
Cruachan	400	7,200	Operational
Foyers	300	6,300	Operational
Dinorwig	1,800	9,100	Operational
Coire Glas	600	35,000	Planning consented
Balmacaan	600	35,000	Proposed
Sloy	60	7,500	Planning consented
Glyn Rhonwy	100	1,200	Proposed
Cruachan 2	400	TBC	Proposed

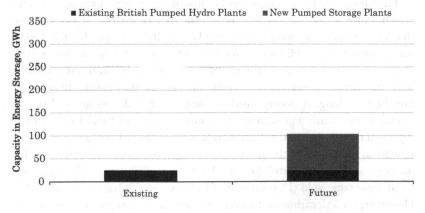

FIGURE 4.5 British energy storage mix if all proposed or planned pumped storage plants are built compared to the existing electricity storage mix.

Feeling the heat: Storing water for flexibility

Pumping water up mountains is not the only way of bringing hydrology into the energy storage mix. Much like batteries store electricity for when it is needed, a hot water tank stores heat to be used on demand. Heating a 150-litre water tank from 20°C to 60°C takes more than 7 kWh of electricity, an energy content equivalent to running a 50-W light bulb for 140 hours.

In some countries, including the UK, controlling when hot water tanks heat up is a means of adding flexibility to the electricity system. In Britain, cheap overnight tariffs are available in some areas to encourage water heating during the night when the electricity networks are less heavily loaded. In New Zealand, a home occupant pays less for hot water if they allow the local utility company to remotely switch off their hot water heaters when the electricity grid is under stress. The home still gets hot water, but the tanks are heated when it is easiest for the utility to provide the electricity for doing so.

Switching off hot water tanks is adequate for reducing peak demand of electricity for a few hours. However, in a future electricity system, it would be really useful to use hot water to shift some of our electricity consumption for several days at a time. For example, strong winds could lead to a surge in wind generation which is followed by a dramatic drop in wind generation for days during the "calm after the storm." During windy weather, it would be ideal to store any extra electricity as low-grade heat which can be used instead of electric water heaters when the weather is calmer.

Heat is interesting to engineers because it is relatively easy to store and also because water heating is a significant user of energy; it is estimated that 17% of domestic energy use in the UK in 2013 was attributable to hot water (Department of Energy and Climate Change, 2013), and better ways of providing it has attracted the interest of some of the brightest scientists. The historical challenge of hot water tanks is that they cannot keep water hot for very long. A poorly insulated hot water tank can go cold in just twelve hours, while a good battery would lose around 1% of its stored energy in a month. Inefficient heat storage such as leaky tanks are bad for the climate as they increase the amount of energy used to make hot water. This matters so much that British Standards now mandate the level of insulation on all new tanks sold (Department of Energy and Climate Change, 2013). However, for a decarbonised electricity system, it is necessary to store the heat for several days rather than several hours.

Hot water presents a good example of this. Consumers are not concerned when and how they get hot water, they just want that hot water when they need it. If Britain was to consider domestic hot water as flexible in 50% of the British homes with water tanks, it would add flexibility to the electricity system which is comparable to all the existing and proposed pumped storage installations in the country. There are a number of companies trying to improve the efficiency of hot water, and by doing so, they could change the course of hot water and become pioneers in multi-day flexibility in the British electricity system. Examples include a *heat battery* which stores heat in a low-temperature phase changing material. This has significant advantages over traditional hot water tanks; the heat is stored at a much lower temperature, so it is much less leaky and able to store heat for up to three days; it is much more energy dense and so frees up space in the house; and it comes at a comparable price to hot water tanks. Phase change materials are perfectly placed for this role. They are thermally efficient that they allow several days between making heat for hot water and ultimate consumption, much like refrigerators decouple between when food is purchased and when it is eaten. This flexibility would allow wind from a Monday to be used for heat on a Friday. Phase-changing materials as an alternative to hot water tanks show us that through some clever thinking, it might be possible to increase the efficiency and flexibility of electricity using another energy vector, in this case heat (Figure 4.6).

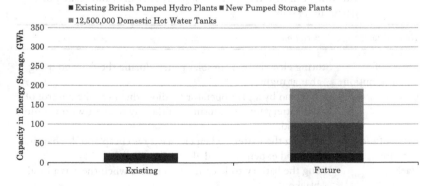

FIGURE 4.6 A future electricity storage mix including pumped storage and flexible hot water.

The domestic store-age

Hot water is not the only way of bringing flexibility from our homes into the electricity system. The battery has always been part of my professional life, from my PhD looking at what they can do for the electricity networks through to working on early solar photovoltaic (PV) and domestic storage propositions in the UK. In 2018, behind the meter storage became the largest market for stationary batteries in some countries and analysis by Navigant Research showed that the residential battery storage market in the USA was expected to grow from $744 m in 2016 to $3,600 m by 2022 (Navigant Research, 2016). A driver for the growth of domestic storage has been co-location of batteries with solar photovoltaic where electricity generated by the solar panels during the day is stored in the battery and used at night. The batteries also have other benefits to the grid which utilities are prepared to pay for, some of which are listed in Table 4.2.

The costs of batteries are falling quickly due to large-scale manufacturing which is causing the huge excitement about the sector. As of March 2018, there were 890,000 domestic solar installations in Britain, nearly three quarters of which were installed over a period of just five years (Department for Business, Energy & Industrial Strategy, 2018). Should the industry grow again at the pace seen from 2012 to 2016, there could be nearly three million solar homes in Britain before 2030.

In the British market, using prices in 2017, the optimal battery size for this application was 6–14 kWh (subject to a few assumptions around the

TABLE 4.2 Domestic batteries provide a number of services to the electricity grid which can help reduce bills, support networks and make power systems more stable

Role of storage	Description
Solar self-consumption	Storing electricity from solar panels during the day and using that at night.
Grid services	Being paid by a grid operator to allow them to take control of the battery for certain times of the day to assist with running the grid.
Time of use tariffs	Charging the battery when electricity is cheap and discharging to avoid expensive grid electricity.
Backup	Using the battery to keep lights running when there is a local blackout.
Fast charging of vehicles	Domestic batteries can work in parallel with the grid to push more power into electric cars for faster charging rates.

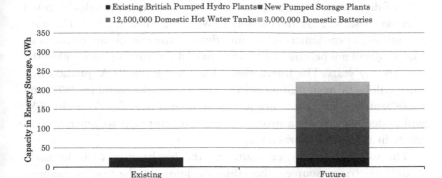

■ Existing British Pumped Hydro Plants ■ New Pumped Storage Plants
■ 12,500,000 Domestic Hot Water Tanks ■ 3,000,000 Domestic Batteries

FIGURE 4.7 A future electricity storage mix including pumped storage and flexible hot water and domestic batteries.

homeowner's behaviour). If 50% of them have battery storage of 10 kWh, then this would add 30 GWh of flexibility to the electricity system (Figure 4.7). It is an aggressive and ambitious scenario for the industry, but it is one that means British homes could start providing flexibility which is comparable to all of the existing pumped hydro resource on the present electrical system. A more thorough examination of what the domestic store-age could mean for homes is provided in Chapter 7.

Electric vehicles and the energy storage mix

In March 2018, there were over thirty-one million cars registered in Great Britain (Department for Transport, 2018). Swapping just some of these over to electric traction will increase our demand for power but also bring new forms of flexibility because cars will not necessarily need to be charged immediately after being plugged in. Electric vehicle batteries need to be large to give them suitable driving ranges: in 2017, the average home battery had a capacity of 6 kWh, while electric vehicles need batteries over ten times that size to have a viable range. Advances in batteries are reducing costs, improving battery life and extending energy density. This has been vital in making electric cars available to hundreds of thousands of people around the world. Once range anxiety is overcome and prices fall, there is evidence that the growth of electric transport could come about much more rapidly than people expect. In November 2017, Matt Finch at the Energy and Climate Intelligence Unit performed a simple projection which compared the increasing sales of alternative fuel vehicles[1] and falling sales of diesel cars. He found that sales of alternative fuel vehicles could overtake diesel by 2021 and possibly by May 2019 (Finch, 2017). There are few utilities or consumers who think that electric cars (using hydrogen or batteries) will not become the main form of domestic transport within the next twenty years. The International Energy Agency (IEA) predicts that there will be thirteen million electric cars around the world by 2020, and those vehicles will provide not just an additional need for electricity. EVs will undoubtedly foster a strong desire from consumers to generate electricity in a clean, low-carbon way.

The value of electric car batteries, the mechanisms to make them flexible electricity sources, the ethics of doing so, the price structures, etc. are all a topic which could fill pages of this book. However, there is a very rational argument to suppose that some of the battery energy storage in vehicles can and should be used, at least in part, to help provide flexibility to the utilities that run a decarbonised electricity system. Suppose that electric cars do begin to rival fossil fuel cars and that just 20% of all of the cars in Britain were electric. If so, suppose that just 30% of those 6.2 million car batteries are available at any one time to offer flexibility to the grid. In such a conservative scenario, an extra 93 GWh of energy storage capacity would be added to the electricity system. This is a flexibility resource that is greater than all of the proposed pumped storage plants in Britain. Even under this conservative estimate, vehicles could add a significant amount of flexibility to help manage the low-carbon electricity system (Figure 4.8).

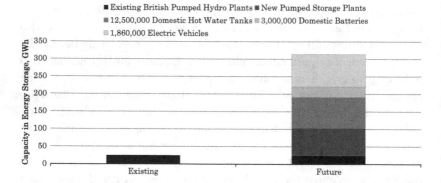

FIGURE 4.8 A future electricity storage mix including pumped storage, hot water, domestic batteries and some of the capacity of electric vehicles available as flexible resources.

Flexibility does not just come from batteries

The energy storage mix is not just limited to our homes and our cars. It is estimated that nearly 100 MW of large battery energy storage was installed in Britain in 2017 such as battery farms tied to solar plants (UK Battery Storage Project Database, 2018). These perform a range of services including fast response to mismatches between supply and demand for electricity, store renewable power generation, reduce consumption of grid electricity at peak times and store electricity when it is cheap.

Such grid-scale batteries and flow battery storage are one such area of additional flexibility, but in the future, flexibility will not just come from batteries as hydrogen storage is being trialled as an alternative to natural gas in heating and in electric cars. Hydrogen can be produced, stored and used at different times and used for both heat and electricity.

In the future, many technologies will have a role to play in British electricity flexibility alongside the snapshot of the energy storage mix identified here. Flexibility can come from technologies or behaviours which alter electricity consumption such as smart appliances which change their electrical load under *demand response*. These devices can be incentivised by utilities through tariffs which provide low price overnight rates. Alternatively, customers might allow their supplier automatically to turn smart appliances on and off to support the grid in return for cheaper power. The focus of this chapter on storage should not diminish the amazing potential of demand side technologies in decarbonisation and power system stability. Traditional forms of flexibility including interconnectors to other electricity grids and generating plants such as suitably designed biomass plants are also likely to have an expanding role in low-carbon electricity. New energy is about adding means of matching supply to demand through new technology, as well as the decarbonisation of electricity generation.

Conclusions

Stores of coal and gas have always been used as a high-carbon means of holding sufficient energy to run the British electricity system for days or months at a time. However, in a low-carbon energy system, the ability to store heaps of coal goes away as soon as coal plants are switched off and another buffer needs to be found. This chapter has reviewed just part of flexibility that could be coming onto the British electricity system as a result of the changing commercial models and technical advances in energy storage.

The volumes of storage proposed are huge in the context of the existing electrical storage; yet the 300 GWh of storage shown for Britain represents less than ten hours of the national electricity demand which is a rapid decline in capacity offered by coal stores. But does that provide enough flexibility for a low-carbon energy system to be built?

In the next chapters, we will review what this electric storage can do alongside a decarbonisation of the national energy mix. This will be done by looking at the role storage is playing around the world as well as developing a vision for lower carbon electricity.

Note

1 Alternative fuel vehicles are fully electric or hybrids.

References

Department for Business, Energy & Industrial Strategy, 2018. *Solar Photovoltaics Deployment*. [Online] Available at: www.gov.uk/government/statistics/solar-photovoltaics-deployment [Accessed 11 08 2018].

Department for Transport, 2018. *Vehicle Licensing Statistics: Quarter 1 (Jan - Mar) 2018*, London: s.n.

Department of Energy and Climate Change, 2013. *United Kingdom Housing Energy Fact File*, London: s.n.

DNVGL, 2016. *The Benefits of Pumped Storage Hydro to the UK*, London: DNVGL.

Finch, M., 2017. *Diesels - Running Out of Gas?* [Online] Available at: https://eciu.net/blog/2017/diesels-running-out-of-gas [Accessed 25 08 2018].

Gridwatch, 2018. *G.B. National Grid Status*. [Online] Available at: www.gridwatch.templar.co.uk/ [Accessed 18 08 2018].

MacKay, D., 2007. *Sustainable Energy Without The Hot Air*. Cambridge: s.n.

Navigant Research, 2016. *Residential Energy Storage*. [Online] Available at: www.navigantresearch.com/reports/residential-energy-storage [Accessed 11 08 2018].

Schulman, D., 2015. *Time to Democratize the Banking System*. [Online] Available at: www.cnbc.com/2015/07/21/paypal-ceo-time-to-democratize-the-banking-system-commentary.html [Accessed 16 04 2019].

The Engineer, 2017. *First New UK Pumped Hydro Scheme for 30 Years Given Go-Ahead*. [Online] Available at: www.theengineer.co.uk/first-new-uk-pumped-hydro-scheme-for-30-years-given-go-ahead/ [Accessed 11 08 2018].

UK Battery Storage Project Database, 2018. *UK Battery Storage Project Database*. [Online] Available at: http://marketresearch.solarmedia.co.uk/reports/uk-battery-storage-project-database-report [Accessed 13 10 2018].

Wilson, I., McGregor, P., & Hall, P., 2010. Energy Storage in the UK Electrical Network: Estimation of the Scale and Review of Technology Options. *Energy Policy*, 38(8), pp. 4099–4106.

5

NEW ENERGY IS VERY DIFFERENT

Thomas Edison reportedly said, *"I'd put my money on the sun and solar energy. What a source of power! I hope we don't have to wait until oil and coal run out before we tackle that."* Sometimes humans take a while to learn the lessons from the past. The world is again looking at solar (and other renewable energy technologies) and in doing so is fundamentally re-evaluating the way energy is delivered.

Electricity grids underwent change in the 2010s, which was unlike anything seen before in the industry and it is likely that the rate of change will only increase in coming years as new technology is adopted more readily. That change not only affects electricity generation, it also affects how consumers engage in energy systems, how decisions are made on the future of power, when and how electricity is consumed and the investments made in electricity by consumers and utilities. Some of these changes are policy changes which rest with governments, while others are social changes that are likely to take place as a result of changing technology. It is important to try to understand these changes to predict what might comprise the future electricity system and whether it is possible to decarbonise it. The next section details some of the key drivers and decisions affecting Britain's low-carbon electricity future and how consumers and utilities might help it to develop.

Power to the people

A low-carbon energy system should not solely rely on a handful of centrally owned and managed power stations. Instead, electricity generation and consumption of tomorrow will consist of millions of devices spread across the grid; large nuclear power stations will work in tandem with small distributed low-carbon generators like wind and solar, and pumped storage facilities of the size of mountains will work with millions of electric car batteries to provide a flexible and responsive grid (Figure 5.1).

A sustainable electricity system is as much about new low-carbon power stations owned by utilities as it is about distributing new electricity assets in communities and homes. New energy decisions will be increasingly made

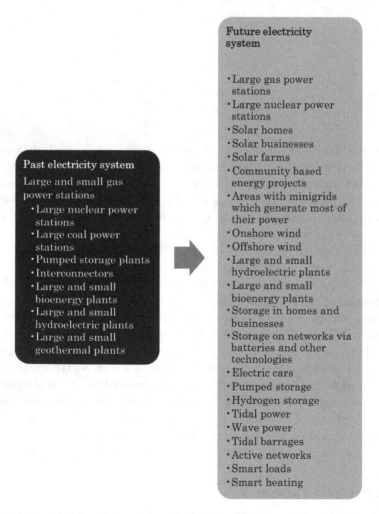

Future electricity system

- Large gas power stations
- Large nuclear power stations
- Solar homes
- Solar businesses
- Solar farms
- Community based energy projects
- Areas with minigrids which generate most of their power
- Onshore wind
- Offshore wind
- Large and small hydroelectric plants
- Large and small bioenergy plants
- Storage in homes and businesses
- Storage on networks via batteries and other technologies
- Electric cars
- Pumped storage
- Hydrogen storage
- Tidal power
- Wave power
- Tidal barrages
- Active networks
- Smart loads
- Smart heating

Past electricity system

Large and small gas power stations
- Large nuclear power stations
- Large coal power stations
- Pumped storage plants
- Interconnectors
- Large and small bioenergy plants
- Large and small hydroelectric plants
- Large and small geothermal plants

FIGURE 5.1 The electricity system of the future will have a larger number and a variety of components than the power system of the past.

by individuals and families to suit their own particular economics, desires and electricity consumption patterns. For example, the rate at which consumers adopt more efficient appliances will not be wholly planned in the corporate offices of energy companies. Instead, this will be brought about through a combination of factors such as regulations, standards, innovation and consumer purchasing power. Homes and businesses could choose to generate their own electricity using solar, wind, hydroelectricity or bioenergy to supplement what they buy from the grid. Consumers could also adopt smart technology and/or energy storage to change when they buy electricity to match when green electricity is being generated on the grid. The power of consumerism does not just extend to generation and storage; energy efficiency and behavioural change are having an impact on decarbonisation. Reducing energy demand can make it easier to build a low-carbon system if it reduces the amount of low-carbon energy sources that need to be found. By opening up more choices, new electricity should place much more control back into the hands of domestic and business consumers, and few technologies illustrate this more clearly than domestic solar panels which have allowed millions of homes around the world to generate some of their own electricity.

Between April 2010 and March 2019, the British Government backed a subsidy programme for small-scale solar photovoltaic (PV) systems (and other low-carbon generators) called the Feed-in-Tariff (FiT). The FiT was a payment for every unit of electricity that solar panels produce over a twenty- to twenty-five-year period, and small domestic solar photovoltaic systems got some of the highest payments. In the early years of the tariff, domestic solar installations earned their investors more than three times the average electricity tariff. The benefits of solar for domestic customers also extended to a reduction in electricity bills as solar power provided an alternative source of energy to the grid during the day. On top of those payments, owners also got paid for solar power that they exported to the grid for the rest of the country to use. When the scheme was announced, the costs of solar meant that the investment case was adequate, but as the costs of solar began to fall returns to investors increased. At times, the FiT scheme was so profitable that in some months, more than 20,000 solar photovoltaic systems were being registered for the subsidy. Over the course of the scheme, the government would cut the subsidy sharply to extend payback periods and to try to subdue install rates. After a sharp subsidy drop, average installations rates could fall five-fold. As solar costs continued to fall, adoption rates would then steadily return as consumers responded to improving financial returns leading to another sharp and sudden subsidy cut. Industry professionals called this the "solar-coaster" in reference to the ups and downs in the market. A similar experience ensued as a result of the mechanism for supporting large-scale solar installations, the Renewables Obligation Certificates (ROCs).

The FiT and ROCs showed the responsiveness of homeowners and businesses to the economics of solar. This is evident in Figure 5.2 which shows the number of domestic solar installations that took place each month in Britain under just the FiT versus the market return at different times. A modest swing in return is seen to add tens of thousands of installations a month across the domestic sector. If there were to be 20,000 installs per month of domestic solar photovoltaic systems over a ten-year period, this could bring more than 8 GW of solar generation to Britain (compared to the 12.6 GW installed between 2011 and 2017 on homes, businesses and in solar farms).

The central planning power of major utilities, which previously determined how electricity is generated, is having to adapt to this new paradigm. Utilities now need to project how fast small generators will be adopted by consumers in terms of how many large power stations are actually needed. Network companies have had to develop new procedures to determine how to connect thousands of low-carbon power stations to the grid and how much to charge for doing so. Power network models have had to adapt to check that electricity can still be supplied reliably. Engineering institutions have had to develop new standards for everything from solar panels to

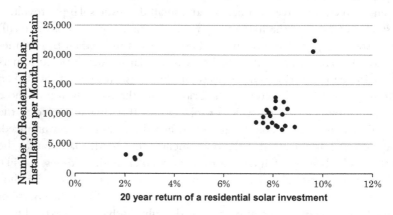

FIGURE 5.2   Number of British residential solar installations from April 2010 to March 2017 when compared to the return on investment.[1] The sector is very sensitive to the rate of return, meaning that when the investment proposition is good, then there should be a higher rate of domestic solar installations in Britain (analysis by author).

switchgear to ensure a safe and high-quality power supply. National Grid and network companies do not decide which homes and businesses install solar panels and how many they should buy. Their role in a decarbonising electricity system is facilitating the connection of new energy to the grid, managing the sustainable transition and keeping the lights on.

Recognising the public value of an energy hierarchy

The politics of future energy are often discussed with reference to the energy trilemma; the importance of an affordable, reliable and sustainable electricity mix (Figure 5.3). These factors can sometimes conflict with each other, for example, when subsidies for low-carbon renewable generation were introduced to boost sustainability, it was recognised that subsidies could increase electricity bills in the short term. The impact of solar and wind on energy security is debatable. Although they produce electricity locally and reduce dependence of foreign fossil fuels, they cannot guarantee to produce power when needed. However, both wind and solar could be considered to be part of a sustainable energy mix: the third part of the energy trilemma.

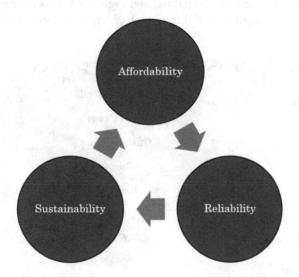

FIGURE 5.3 The energy trilemma is the supposedly competing demands for low-cost energy, a reliable power supply and the provision of energy in a sustainable manner. The ideal energy mix uses low-cost, low-carbon electricity sources and assets to make power systems reliable.

Similarly, gas and coal power stations have historically been viewed as delivering affordable and reliable energy as they have the ability to provide electricity when needed such as on cold, dark winter nights. However, the increasing amount of imported gas and coal is a threat to energy security as foreign energy supplies are more susceptible to the moods of international politics and price fluctuations. Fossil fuels are also, of course, intolerable from any environmental perspective.

There is a general perception among policymakers that the energy trilemma represents what is important to the public in terms of a future energy system. Electricity is important for our health, security and economy, and it should, therefore, be more affordable, reliable and sustainable. However, with gas and coal not fulfilling the sustainability argument and their long-term reliability in doubt, a more viable energy mix needs to be found for Britain.

A sustainable energy mix needs to increasingly focus on what is important to the people who have to pay for and rely on it. That mix will contain at least all of the low-carbon generation forms, but the proportion of wind, solar, nuclear, biomass, tidal, hydro, etc. is up for debate. A barometer for judging a future energy system might also be the acceptance of different technologies by the public, and Figure 5.4 shows the results of a survey into the popularity of different electricity generation technologies.

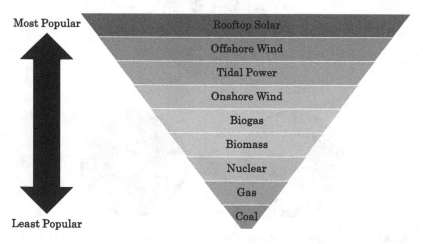

FIGURE 5.4   Acceptance of different energy sources from the most popular (solar) through to the least popular (coal) shows how low-carbon energy is favoured among those surveyed (ClientEarth/YouGov, 2018).

Rooftop solar, offshore wind and tidal power top the list, while gas and coal are some of the least popular. Nuclear power is the third least favourite technology, presumably due to the perceived safety/environmental concerns with the technology and the high costs associated with recent nuclear power stations. In line with this particular survey and climate goals, it is irrefutable that the role of gas should be kept to a minimum, and coal has no future. Coal power may have powered the nation for over a century and provided millions of jobs, but it is high-carbon, mostly imported and does not help provide sustainability. The energy security of a finite fuel is also high circumspect.

Energy is inherently political, and energy policy has historically been determined at a national level. However, there is evidence of increasing consumer power in the decisions surrounding the future electricity mix. In this hierarchy, renewable energy from solar, wind and tidal would take precedence over nuclear power and biomass. Yet there are many benefits from a diverse energy mix which includes the nuclear power favoured by some actors in the electricity industry. Balancing those such conflicting arguments through rational and informed debate is key to creating a fair, affordable and environmentally responsible power system.

The role for nuclear power and challenging old investment models

In developing our low-carbon toolkit, it is prudent to recognise that the power of the people might have a tremendous impact on changing the way a decarbonised electricity system could be built. In addition, the increased feasibility of renewable electricity generators is also threatening established low-carbon technologies. Nuclear power, which has a very low-carbon factor is a clear case in point; nuclear is presently proving to be expensive, highly controversial, potentially dangerous and results in radioactive waste deposits which need to be protected for thousands of years. Guaranteeing political stability and a faultless system for managing nuclear waste for tens of times longer time than humans have had mechanised transport rightly concerns people! Yet from an engineering perspective, nuclear power stations are very appealing because they generate a predictable baseload of low-carbon energy and operate in a way which complements how the power grid presently operates.

It is important to keep an open mind to the future of nuclear power in the electricity mix, as with any other technology. In Britain, there is sufficient power demand to justify adding further nuclear stations, while electric vehicles and electrification of heat should raise electricity consumption and permit even more nuclear power plants than can be justified today. However, meeting all of the present electricity requirements only from nuclear plants is probably impossible. To generate the nearly 300 TWh of electricity that was consumed in 2016, Britain would need more than ten new nuclear power stations each the size of Hinkley Point C running continually. If these all had the same supply contract as that new nuclear power station, electricity prices will rise. Nuclear power is not cheap to build, run or decommission and low-carbon alternatives, can already produce energy at comparable or lower prices and their costs are declining every month (Figure 5.5).

In addition to economic challenges, nuclear power has serious technical constraints which prevent it being the sole technology to bring about decarbonisation. Nuclear power stations are not readily designed to be switched on and off. If the amount of nuclear power exceeds the demand for electricity, then some means to consume all of that electricity generation needs to be found. As a result, in a fully nuclear power system, the timing of when energy is consumed needs to change, with much more electricity consumed overnight to flatten demand, including the coldest

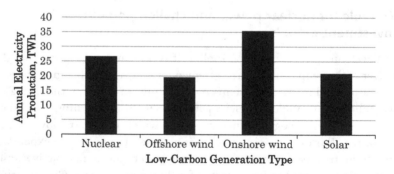

FIGURE 5.5 Annual electricity production from nuclear power plant at Hinkley Point C (at 97.5% availability) versus an equivalent £18 billion investment in offshore wind (5.8 GW with capacity factor 38%), onshore wind (15 GW with capacity factor of 27%) and solar (22 GWp with a yield of 950 kWh/kWp). Analysis by author using data from Bloomberg New Energy Finance (2016) and Renewable UK (2018). This demonstrates how the cost of Hinkley Point C can be compared to an investment in low-carbon alternatives when it was announced. Wind and solar prices have dropped significantly since that time.

days when demand is usually high to the warmest summers when consumption traditionally falls. Accordingly, Britain will need investment in flexible electricity technologies to smooth demand and backup the nuclear power plants; a criticism which can equally be levelled on variable renewables which need storage to match supply to demand when the wind is not blowing and the sun is not shining strongly enough to make sufficient electricity. Due to economic and technical constraints, in no system should nuclear energy be the sole or major component of low-carbon energy.

New energy

The national electricity system of today actually began as separate networks located in towns and cities across the country. Today, these might be referred to as a set of microgrids. Eventually, these microgrids were joined together into a single network distributing power from large central power stations to loads all over the country under the management of a central electricity authority. From that moment on, electricity ran as centrally planned and managed system and remained so until it was privatised by the government in the 1990s in an attempt to create a competitive marketplace, keep costs low and to drive innovation. Central planning functions in the grids mostly remained, with the old management structures reflected in the newly privatised companies.

In the absence of true competition, it is difficult to argue that the early years of privatisation did little other than hand control of the national electricity system from ownership of the people to the ownership of private corporations. It has widely been recognised that the privatised electricity system initially came with flaws that conflicted with the decarbonisation agenda. The electricity system, even in private hands, remained a collection of a small number of actors running large power stations and networks and one centrally managed system operator. Privatisation initially resulted in a small number of corporations whose market share regulators would continually try to contain, particularly when takeovers led to the risk of larger monopolies. Such monopolies have very little incentive to innovate when the status quo supports their way of operating, which is particularly true of networks who until recently have had no competition at all for local power supply. Suppliers and generators grow their businesses as demand increases which can conflict with a key means of reducing fossil fuel use (reducing consumption) or they can reduce costs which itself can lead to a decline in service levels. However, in recent years, the private sector and the way it is regulated has evolved to some degree which hints at a better mode of operation to bring about low-carbon electricity. Innovative suppliers and generators have come to the market to create competing products for consumers such as those which only offer low-carbon electricity. Monopolies which run British power networks are now allowed to generate income from providing reliable and efficient networks and for innovating with new technology to reduce costs to consumers.

Electricity systems have always evolved by consulting a precisely defined and well-known set of experts and models and the privatised markets initially concentrated on building large generating plants under this way of working. Until recently, power stations were added to the system at a slow rate. This legacy means that companies can be slow to respond when the rules fundamentally change. As an example, the ability to stream movies completely changed the way in which films are viewed at home and ended the viability of video rental firms. The rapid uptake of renewables in the 21st century has shown how the electricity system might face the same fundamental change.

The FiT and ROC has allowed a competitive industry to develop which has deployed wind farms, solar farms, hydro plants and bioenergy across the country. It has created a marketplace for millions to participate in the electricity system to a greater degree than ever before. It shifted the paradigm from a central planning team which could make all of the decisions to one which now needs to learn to respond to rapid and unpredictable changes which are partly beyond its control. In the modern energy system, consumers start to become a much more powerful force and can generate the own electricity rather than buying it from the grid. As a result, the levers that planners can use to control the evolution of the power system have shifted from deciding which power stations to build to inventing incentives and regulations to steer consumer behaviour. In addition, new technology and commercial realities mean that other forms of generation are now competitive.

The lights have stayed on, and by all appearances, the central planning of the electricity grid has responded well to the change. However, the rules under which they operate have fundamentally changed and that needs to be reflected when considering what the future electricity system might look like.

Conclusions

There should be little debate whether Britain should aim to create a more socially, economically and environmentally responsible low-carbon energy mix. Perversely, however, there remains much debate about whether the present electrical evolution underway in Britain is able to achieve those aims. Traditional approaches to electricity are changing, and there are now methods for people and utilities to engage with the future of the power system in ways which were unimaginable at the turn of the millennium. The power of the consumer is now beginning to challenge the established control of electricity companies and governments who are used to levering the energy system.

Looking ahead for Britain, it is reasonable to question whether a key and non-negotiable aim of electricity can ever be achieved: that of lower carbon energy. In the next chapter, a vision of various actors in energy is presented and simulated to determine whether lower carbon electricity is ever achievable. From that position, it is possible to address the social and economic potential of future electricity.

Note

1 Rate of return calculated from subsidy, electricity savings and export tariffs. Highest install rates not shown.

References

Bloomberg New Energy Finance, 2016. *Offshore Wind Could Replace Hinkley in U.K. at Same Cost.* [Online] Available at: www.bloomberg.com/news/articles/2016-08-16/offshore-wind-could-replace-hinkley-nuclear-in-u-k-at-same-cost [Accessed 06 02 2019].

ClientEarth/YouGov, 2018. *ClientEarth's Climate Snapshot: A Survey of UK Attitudes Towards Climate Change and Its Impacts,* London: ClientEarth.

Renewable UK, 2018. *Wind Energy Statistics Explained.* [Online] Available at: www.renewableuk.com/page/UKWEDExplained [Accessed 01 09 2018].

6

A VISION FOR DECARBONISING BRITISH ELECTRICITY

Caroline Lucas is a Green Party politician from the UK. In 2015, she brilliantly summarised one of the most striking challenges about the British decarbonisation strategy when she said,

> This switch from coal to gas is like trying to go dry by switching from vodka to super-strength cider – it entirely fails to seriously address the real challenge at hand. Investing in renewables and energy conservation would be far more effective economically, environmentally and in terms of energy security. We must begin weaning ourselves off gas as quickly as possible.
>
> *(Lucas, 2015)*

This chapter is about assessing if that statement is at all possible – can Britain switch from coal and gas onto a much more decarbonised electricity mix?

A vision for lower carbon electricity in Britain

As per the energy trilemma, changing electricity has value if it improves sustainability, affordability and reliability. Chapter 3 evaluated the impact of three changes on the sustainability of British electricity over a single month. In this chapter, a vision for what decarbonised electricity might look like in Britain by 2030 is developed to reflect the fundamental ways that energy is changing. Using some of the techniques I apply in my professional life for modelling global power supplies, the vision for lower carbon electricity is assessed over every hour of every day for an entire year. Knowing that decarbonisation has to happen, and with credible evidence from power system data about what a low-carbon toolkit can do, the vision for lower carbon electricity is about seeing what is achievable over the next decade to meet 2030 interim carbon goals.

Electricity demand

There is a large degree of uncertainty about what Britain's electricity requirements might be in 2030. Historical trends in the consumption of electricity are a notoriously unreliable basis for long-term energy forecasts, as consumption is affected by a variety of parameters such as the population growth, changing economies, energy efficiency and changes in electrical technology. As identified in Chapter 2, British electricity demand actually fell between 2010 and 2018 in part as a result of improving energy efficiency of lighting, appliances and rising electricity prices. Various studies seek to project future electricity demand, and there is a large amount of disparity between these. Some experts predict that demand will also increase as a result of increased industrial activity, electrification of heat and transport, etc.

Regardless of what happens to heat and transport, which are not included in this vision,[1] it is fair to say that Britain's policymakers probably want improvements in efficiency to continue to reduce the demands of the existing uses of electricity on the grid. Increased efficiency means Britain can use electricity for the same reasons, but with less of an impact on the climate and, if done properly, for less cost to consumers. Excluding needs for transport and heating, Figure 6.1 shows scenarios for how electricity demand might change between 2018 and 2030. One scenario sees demand continuing to fall at the rates seen between 2008 and 2017. This is unlikely as efficiency gains seen in recent years and reducing demand would slowly get less effective as the "low hanging fruit" of inefficient products get replaced. For this scenario to be credible, efficiency gains would need to be sustained and expanded to other present electricity uses such as heating, computing and industrial machines.

The vision for lower carbon electricity includes what should and could happen to reduce carbon and energy bills: that being demand reduction through efficiency. This closely mirrors National Grid's highest demand reduction scenario in the 2018 Future Energy Scenarios report

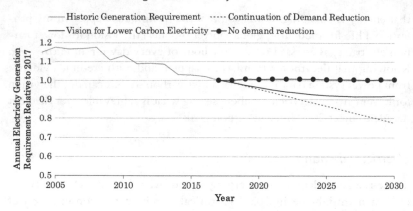

FIGURE 6.1 Various scenarios for changing electricity demand in Britain including the scenario used in the vision for lower carbon electricity. Note that some solar, wind and hydroelectricity is charged into and subsequently discharged and is shown as being from storage.

(National Grid, 2018) and sees electricity consumption falling to just over 90% of 2017 levels. Further demand reduction might be found beyond what is assessed in the vision, and that can be a good thing for decarbonisation if it is done correctly. By setting a higher demand target, the vision sets a harder target for decarbonised electricity generation.

Wind generation

The role of variable renewables in future electricity is widely debated for economic, environmental, social and technical grounds, yet there is little debate about whether they will have a place of some form in the future electricity mix. Wind power has undergone sustained growth since the dawn of the renewable electricity age. Over 19 GW[2] of turbines were installed in Britain by the end of 2017, which produced more than 10% of British electricity (Department for Business, Energy and Industrial Strategy, 2019; MyGridGB, 2018). An increasing amount of this wind capacity will be installed offshore, where wind speeds are generally higher and more consistent, while manufacturing techniques have made wind technology more efficient. For this reason, wind turbines of the future should produce more energy than early technologies. Various projections for the future of wind are shown in Figure 6.2. If the increase in wind capacity was to continue

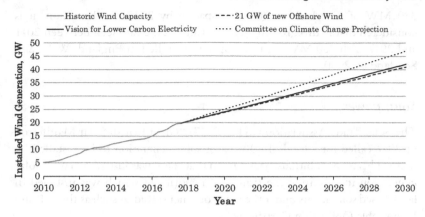

FIGURE 6.2 Historical and projected install rates of wind turbines including the scenario used in the vision for lower carbon electricity.

at the historical rate, then Britain could see nearly 41 GW of wind turbines installed by 2030. The trade body which represents wind generation, Renewable UK, has said that 21 GW of this could be offshore wind turbines installed under existing and future auctions by 2030, a trend that closely matches historic growth (Renewable UK, 2018). Wind has historically outperformed expectations as turbine efficiencies increase and costs, and the Committee on Climate Change (CCC, 2015) "high renewables" scenario of 2015 sees 47 GW of wind by 2030. However, it should also be recognised that it is a market which could eventually saturate due to a lack of suitable sites for turbines. For the purposes of the vision for lower carbon electricity, an average between the projected wind growth seen through historical data and through future auctions is applied.

Hydroelectricity

At the end of 2017, Britain had 1,900 MW of hydroelectric plants, and these provided a very small fraction of annul electricity generation. Despite incentives such as the Feed-in-Tariff (FiT), hydroelectricity has seen much lower growth than other low-carbon technologies (Department for Business, Energy and Industrial Strategy, 2019) and due to British topography hydro will never be a major contributor to British electricity. However, as a low-carbon power source, hydropower potential could and should continue to be explored where economically and technically viable. Should recent growth in hydropower continue, Britain would see

330 MW of new hydroelectricity capacity by 2030. In the vision, it is considered that hydro growth rates continue at the rate seen between 2011 and 2017 with 360 MW[3] of new capacity installed in England, Wales and Scotland by 2030.

Solar power

The Solar Trade Association's (STA, 2017) "Great British Solar Manifesto" of 2017 called for a total of 40 GW of solar generation to be installed by 2030. This is an ambitious target given that there was just 12.6 GW of solar generation in Britain at the time, the majority of which was installed in land-based solar farms under the since defunct subsidies such as the FiT and Renewable Obligation Certificates.

Under the "power to the people" scenario described in Chapter 5, it is the public as well as utilities that build the low-carbon energy system. This factor is recognised in the National Grid 2018 Future Energy Scenarios report (National Grid, 2018) with a rapid update of solar by homes and businesses. As such, the future should be much more about solar installed on homes and businesses than in fields. If *power to the people* is a credible factor in future energy, it would take more than seven million homes to adopt solar to meet the STA manifesto target, and for these to be installed at an installation rate never seen in Britain.

On the basis of this evidence, the STA target looks difficult to achieve without a return to solar farms. However, a few factors might cause solar to exceed past performance in rooftop and solar farm sectors, higher efficiency panels are meaning that more power can be produced per square metre of roof space and fewer solar installations will be needed to meet targets, declining costs of solar are being compounded by higher electricity pricing making photovoltaics more compelling in all sectors of the economy which will boost the solar industry and cost reduction and new business models, such as lease schemes, are reducing barriers to entry. This is compounded by declining costs of energy storage, which from my own personal and professional experience can make solar more compelling for domestic and industrial consumers.[4] The vision for lower carbon electricity assesses production from solar growing through the power of consumerism between the present installation rate and that required to achieve the Great British Solar Manifesto (Figure 6.3). This is a similar solar capacity assessed in the "Community Renewables" scenario, in the 2018 Future Energy Scenarios report (National Grid, 2018). It is very likely that solar will exceed these forecasts because the market is very

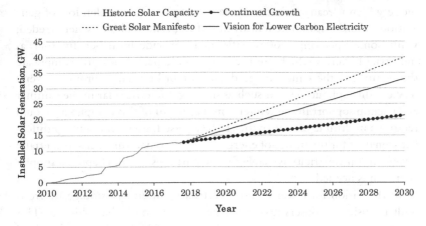

FIGURE 6.3 Various projections for the future of the British solar industry including the scenario used in the vision for lower carbon electricity.

sensitive to economic returns as the economics of solar and energy storage get stronger and because National Grid usually underestimate the growth of the solar industry.

Electricity storage mix

As identified in Chapter 4, there should be many options to increase in the amount of flexible energy storage coming to Britain through decarbonisation, new business models and potentially through the build out of new pumped storage plants. However, the actual likely future of electricity storage in Britain is hard to predict both in terms of the amount of storage that will be installed and the function of that storage to support the grid. To some degree, the future depends on the adoption and commercial viability of large energy storage plants in the form of pumped storage or other large-scale technologies like fields of lithium or flow batteries for example. To another degree, it depends on the embracing of batteries and/or hot water storage in homes and businesses, the adoption of electric vehicles and the willingness of manufacturers and customers to let those vehicles support the grid. At the same time, the operating mode of energy storage will have a huge impact on what it can achieve in terms of decarbonisation; one operating mode might be the classic energy shifting role of making renewable energy available when needed, e.g. shifting solar power from day to night; other operating modes will keep storage in a standby mode which is ready to back up the grid if there is a sudden loss of generation or load with no benefit from shifting renewables to when needed; while some operating modes will simply provide local benefits such as industrial batteries which are used to provide peak loads. To complicate things further, the same storage device might be used to provide several services throughout the year such as shifting domestic solar in the summer and backup support in the winter. Simulating all of this complexity would require a multitude of sensitivities to be assessed, each representing different scenarios for the future of energy storage. To determine if substantially lower carbon electricity is viable for the purposes of this chapter, a simpler approach is needed.

In the vision for lower carbon electricity, an energy storage future is built up using a conservative estimate of what might be achieved. This comprises a mixture of new and existing pumped hydro plants with a blend of utility scale and behind the meter battery storage (installed primarily with the solar power identified previously) as summarised in Figure 6.4. This is around 30% of the total storage capacity identified in Chapter 4 and is a snapshot of the storage that could be brought online by 2030. The vision sticks with simple storage devices which are controlled to charge whenever low-carbon generators on the grid are producing more electricity than demand and discharged as needed by the grid; an operating mode known as *energy shifting*. To be commercially viable, the storage might be needed for other roles in the year, but it is fair to assume that the storage can be made available for energy shifting as required by the grid. Electric

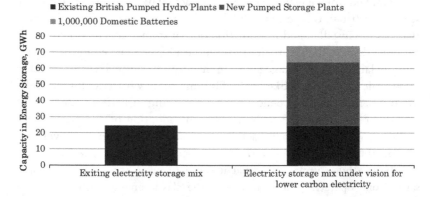

FIGURE 6.4 Electricity storage mix applied in the vision for lower carbon electricity.

vehicles are not included – this would not be fair as the vision is not looking at how to provide low-carbon electricity for those vehicles. Similarly, domestic hot water storage facilities are also not included as these do not store and release electricity into the grid; instead, these produce heat for taps and showers.

This snapshot of the future storage mix assessed in the vision for low-carbon electricity is designed to provide some insight into the potential for electricity systems which can respond flexibly to more variable demand and generation. Although this sees electrical energy storage reserves growing by over three times, a key assessment of the vision will be how frequently these storage devices are actually used for *energy shifting* and the impact this has on decarbonisation.

Bioenergy

The CCC projected limited growth in bioenergy power stations in Britain beyond 2020 under a "High Renewables" scenario for the power system. This was amid concerns that growth in biomass would be unsupported by government due to concerns around sustainability (CCC, 2015). However, by 2017, the former coal power station at Drax, Britain's largest bioenergy power plant, could generate 2.6 GW from biomass and with a potential generate a further 1.3 GW if more of the former coal plant was converted to biomass. Hundreds of smaller bioenergy facilities, such as energy from waste plants, have also been built in Britain over the past decade. If these smaller power stations continue to be built at the same rate, such "distributed bioenergy" alone could exceed the capacity of Drax by 2030.

Could this be sustainable? More bioenergy power stations in Britain does not necessarily mean a directly proportional increase in the volume of biofuels that are burned each year. For example, if the bioenergy capacity in Britain is doubled from 2017 levels but these power plants were used half as often, then the amount of biofuels used would be more or less the same as before.[5] As bioenergy generators are one of the last to dispatch before gas in the vision for lower carbon electricity, it is credible to increase the power of bioenergy plants, without necessarily seeing a proportional increase in biofuel use. In the vision for lower carbon electricity, 12 GW of biomass is used, which is double of biomass capacity in Britain in 2017 and slightly lower than historical growth rates (Figure 6.5). The impact of this on total biofuel use and sustainability use will form a key assessment of the vision.

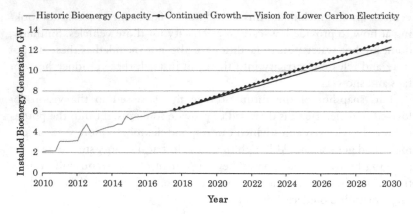

FIGURE 6.5 Various projections for the growth of the British bioenergy industry including the scenario used in the vision for lower carbon electricity.

Imports, exports and electricity interconnectors

Interconnectors are being developed or operated between the British electricity system to other parts of Europe including Norway, France, Belgium, Ireland, Denmark and Iceland for the import and export of electricity. On a simplistic level, exports might occur when there is a surplus of electricity in Britain while imports might occur if there is a shortfall of electricity in Britain or if imported electricity is cheaper than domestic power stations.

In 2017, Britain was a net importer of electricity. How and when interconnectors are used and how much energy is imported and exported through them needs to be re-evaluated in a fundamentally different electricity system. For example, growth in low-carbon electricity in Britain could mean that there could be more surplus electricity generation to be sold to other European countries. The vision, therefore, includes the existing and proposed interconnectors shown in Figure 3.8 but with a restriction that only low-carbon electricity is imported through these and those imports only occur when there is insufficient low-carbon generation in Britain. This restriction in the vision aims to reduce carbon in Britain and Europe.

Nuclear power

Nuclear power stations provide baseload, low-carbon electricity yet it sits low in surveys of public acceptance. The nuclear power station under construction at Hinkley Point in Somerset has shown that nuclear energy can be expensive, even when compared to the costs of managing the intermittency of solar and wind. Despite that, it is highly probable that nuclear power stations will comprise some of the future electricity mix, but the amount built should only be what is needed sufficient to decarbonise when the value of more popular and affordable low-carbon electricity generation technologies are considered. After some optimisation to trade-off between variable renewable production and carbon objectives, the vision for lower carbon electricity includes 11.5 GW of nuclear plants, which is an increase on the 9.5 GW of nuclear plants in operation at the end of 2016 but less than the 16.4 GW of planned/under construction nuclear power stations in Britain.

Thermal power stations

Fossil fuel power stations should have no place in a future energy system which aims to be fully decarbonised. In the vision for lower carbon electricity, there are no coal power stations due to these having the highest carbon emissions of any source of electricity and because these are due to be phased out by 2025 under UK Government policy. However, gas power stations, as the lowest carbon mainstream fossil fuel, are included as a last resort backup and are only allowed to be used when there is no other electricity generation available. Reducing gas from baseload to backup should have a substantive impact on carbon emissions and a key metric of the vision will be whether this is sufficient to meet carbon goals. Further, assessing the volume of gas used when all of the other low-carbon generation is added to British electricity under the vision should provide insight to help investigations into where this gas might come from.

A viable electricity mix?

The CCC Fifth Carbon Budget has also provided scenarios for volumes of generation from different technologies. The "high renewables" scenario considers 47 GW of wind generation,[6] 40 GW of solar generation and 1.7 GW of hydroelectric generation by 2030 in order to achieve a carbon emission of around 100 $gCO_2eq./kWh$ by 2030 (CCC, 2015). The capacities for wind, solar and hydroelectricity in the vision for low-carbon electricity fall below that assessed by the CCC under their most renewable scenario. The CCC "high nuclear" scenario has a comparable amount of nuclear power (11 GW) to the 11.5 GW assessed in the vision for lower carbon electricity. It is worth noting that the CCC also include 1 GW of

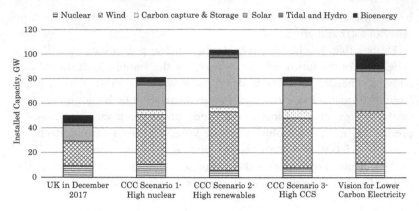

FIGURE 6.6 Installed capacity of lower carbon electricity generation in the UK, under different CCC scenarios and under the vision for lower carbon electricity. Note that (a) carbon capture and storage could be included in the vision and co-installed with generators such as bioenergy and (b) that tidal sources are only not included in the vision due to a lack of data available to the author about their electricity production.

tidal generation which is not included in the vision for reasons of simplicity but could be in the future. The model presented here is not necessarily what I believe as an author will happen or should happen but is a scenario that is most easily defended for the purposes of assessing if carbon targets are achievable (Figures 6.6 and 6.7).

Running the electricity system

If Britain is to implement a vision for lower carbon electricity, a key question for the grid operators is when and how frequently large power plants should be used to meet our electricity demand. Determining when to switch on various plants, the so-called dispatch of generation, is presently determined using a series of market mechanisms under a highly regulated set of rules, yet in a changing energy paradigm, it is unrealistic to expect the same markets rules to remain in place. In assessing whether the vision works from energy balancing[7] and decarbonisation perspectives, an alternative dispatch is implemented. To be useful, this should be a sufficient approximation of the most economic and low-carbon dispatch strategies and should closely reflect the technical constraints of various plants every hour of each day.

In the dispatch strategy used to assess the vision for lower carbon electricity, nuclear plants are always run first to provide baseload electricity.

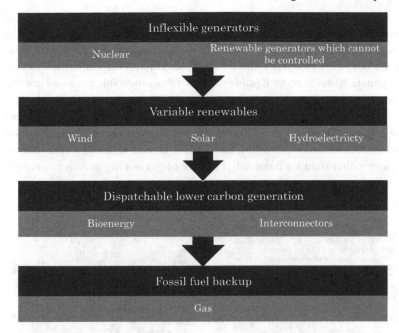

FIGURE 6.7 A vision for delivering lower carbon electricity in Great Britain by 2030. Other interventions such as tidal energy and carbon capture and storage could also be assessed under this framework if desired; these are excluded in this vision for simplicity of reading rather than as a judgment of the value these might have in future electricity.

Although nuclear power is less popular with the public than renewables, technical and economic realities mean that, at present, any nuclear plants that are built should run continually and, therefore, must be run first. After the nuclear plants, variable renewable generators (wind, solar and hydro) are allowed to generate. Any excess electricity from these plants, termed *overgeneration*, is taken into electrical energy storage facilities unless these cannot take any more power or are full. If there is insufficient demand or storage for the variable renewable plants, then it is assumed possible to switch off enough of these renewable plants[8] in order to keep supply and demand in balance. It is also possible for some of this renewable generation to be exported, via interconnectors, to neighbouring nations should these countries have a shortfall in low-carbon power.

If there is not enough wind, solar, hydro or nuclear power to meet electricity demand at a particular time, then energy storage plants are brought online. These will be designed to store low-carbon electricity and be ready to come online to meet demand when required. Storage is used until it is empty or variable renewable generation increases above demand. If the renewables and storage are unable to meet demand on a particular hour of the day, then bioenergy and low-carbon imports are dispatched. These are generally much more flexible than variable renewable plants or nuclear power. Lastly, gas is dispatched as a critical backup. Gas is only used when there is insufficient wind, solar, hydro, nuclear, storage, imports or bioenergy to keep the lights on. This means that an unsustainable, high-carbon and environmentally damaging fossil fuel is used as little as possible and for backup rather than for baseload. The role of gas is a key means to critique the vision for lower carbon electricity (Figure 6.8).

Variable renewables

- Growth of low-carbon renewable generation from wind, solar and hydro enabled by thousands of new and existing actors in the electricity system.
- Other forms of renewable – tidal, wave, tidal stream etc. – not modelled due to a lack of data. Government to support these technologies to further decarbonise energy system.

Demand reduction through efficiency

- Continued reduction in electricity demand as a result of energy efficiency measures.
- Energy efficiency standards reinforced through legislation.
- Recognition of the cost barriers to efficient products for the poorest in society by policymakers and businesses.
- Measures to increase access to safe and efficient products are enacted.

Energy storage mix

- A sensible energy storage mix based on the technologies described in Chapter 5.
- Electric vehicles are not included in the energy storage mix as doing so would also have impacts on demand which are not being modelled here.

New Nuclear

- New nuclear is permitted if it meets strict cost targets.
- Nuclear viewed as a secondary investment to renewables, flexibility and energy efficiency to reflect its status on the energy mix hierarchy.

Biogas, Gas and Coal

- Thermal plant viewed as a backup when storage plants are empty and there is insufficient variable renewable plant to meet demand.
- Biomass and biogas prioritised over natural gas.
- An end to coal use in the UK.
- Carbon prices used to encourage a shift from natural gas to alternative dispatchable low-carbon electricity generators e.g. hydrogen and bioenergy.
- Use of waste heat from these plants in community heating schemes.

People power

- Policies to encourage "behind the meter" electricity generation such as rooftop solar, small hydro and small scale wind.
- Policies including better use of data to help customers become more energy efficient.
- Behind the meter generation and storage used by councils to reduce fuel poverty.
- New business models found to make new energy available to more people.

FIGURE 6.8 Dispatch order according to the model used to assess how low-carbon electricity might be achieved. This is a summary of that shown in Figure 3.10.

Evaluation of the low-carbon electricity vision

The vision for lower carbon electricity has been assessed using weather and electricity demand data from 2017. A comparison of the electricity mix in that year relative to the vision for lower carbon electricity is shown in Figure 6.9. What makes the vision so powerful is a huge increase in the low-carbon electricity mix; however, it is not able to transition completely away from fossil fuels. Nuclear electricity increases its role as a baseload

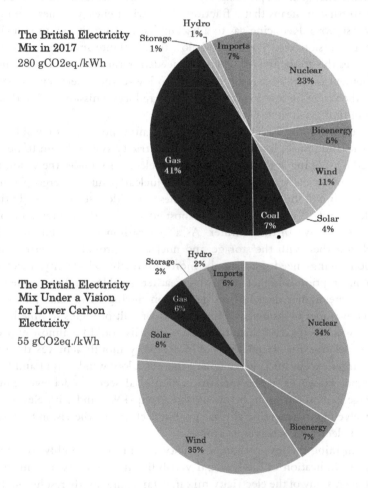

FIGURE 6.9 (Top) the British electricity mix in 2017 and (bottom) electricity supply mix under a vision for lower carbon electricity. This shows what proportion of electricity is supplied by different sources at any particular time.

electricity generator, while wind and solar generation grow to provide a mix of low-carbon electricity sources. Imports and the use of biomass increase, but their use is constrained to times of low wind and sun rather than as regular electricity generators. The role of electrical energy storage is increased, and its main role in the electricity mix is the offsetting of gas by taking surplus electricity from renewables when the wind is strong or the sun bright and moving that to times of adverse weather. The most dramatic change in supply comes from gas, whose changed role from base-load to backup means that a fraction of electrical energy comes from gas power stations. Reducing gas to a backup has a major impact in reducing carbon emissions over the year to below the 2030 interim carbon goal and preserves this finite fuel for when it is needed, rather than when it is most convenient to use. As a result of a shift to low-carbon electricity genera-tion, the vision for lower carbon electricity reduces emissions to less than a fifth of those seen in 2017.

The reasons for this changing electricity mix, how this is possible and what it could mean for a decarbonised electricity system begin to be re-vealed by looking at the monthly share of electricity under the vision for lower carbon electricity (Figure 6.10). Nuclear plants undergo planned and unplanned shutdown events, yet these provide a steady contribution of electricity over the year. Wind is most effective in the winter, and solar is most effective in the summer. As a low-carbon mix, wind and solar work together with the storage and nuclear to provide the year-round carbon savings needed to supply more sustainable electricity. Electric-ity consumption is higher when the weather is colder, and consequently, there were higher demands on generation during these times, some of which was met by using gas. However, as a result of high sunshine hours and lower electricity demand, summer months could be almost entirely fossil fuel free for electricity provision. Every month achieves the car-bon target except for January where there is a low wind output and high demand. However, if the interim carbon goal seeks to achieve annual average carbon emissions below 100 $gCO_2eq./kWh$ and with eleven out of twelve months meeting climate goals it is clear that the vision for lower carbon electricity achieves its aim.

Examining the low-carbon electricity system on a monthly basis pro-vides an indication of the seasonal variability in electricity as simulated. However, study of the electricity mix in detail using hourly resolution data is key to understanding how viable the electricity generation mix might be. Figures 6.11–6.14 show high-resolution charts of the projected hourly elec-tricity mix for four months of the year. This reveals an electricity system

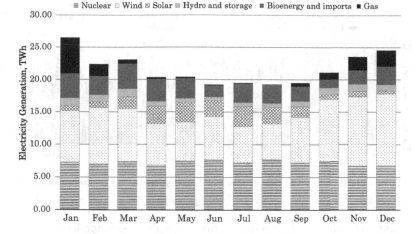

FIGURE 6.10   Monthly electricity generation from different sources under the vision for lower carbon electricity. This shows a consistent contribution of nuclear power, supplemented by wind, solar, hydroelectricity, biomass and imports. The vision still requires gas to meet higher winter electricity demand and raises important questions about gas alternatives. A key question raised by the vision for lower carbon electricity is whether gas can partly or wholly offset using measures such as higher investment in low-carbon electricity generators, seasonal storage of biogas or other similar measures.

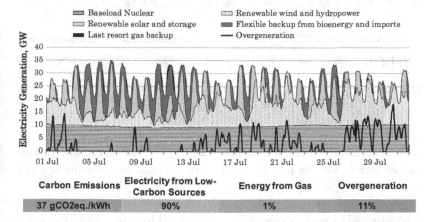

FIGURE 6.11   The hourly electricity mix under the vision for lower carbon electricity if it has been in place in July 2017.

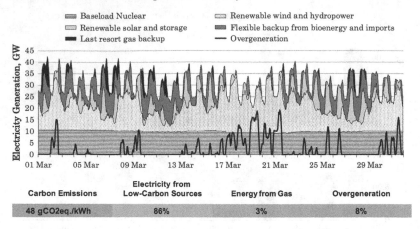

FIGURE 6.12 The hourly electricity mix under the vision for lower carbon electricity if it has been in place in March 2017.

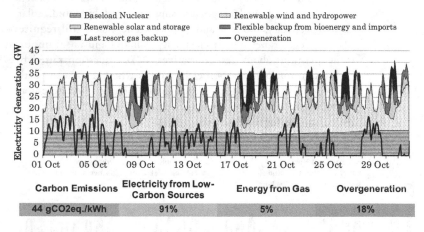

FIGURE 6.13 The hourly electricity mix under the vision for lower carbon electricity if it has been in place in October 2017.

with generation and consumption variability, but one where low-carbon power sources are able to meet most of the national power demand and a flexible backup to keep the lights on.

Figure 6.12 shows one of the lowest carbon months for electricity: July. The baseload of nuclear electricity is supported by wind, solar and some hydro to produce very low average carbon intensity of below 40 gCO₂eq./kWh over the month. Solar generates in the daytime when demand is

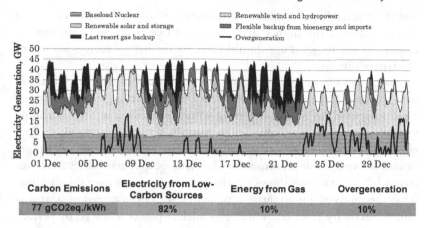

FIGURE 6.14 The hourly electricity mix under the vision for lower carbon electricity if it has been in place in December 2018.

highest, with some of that solar electricity being stored in battery storage to be used in the evenings. The high solar generation means that for some periods such as 18 July, there is enough generation to generate all of British electricity using solar, wind, hydro and nuclear power alone. When the solar and wind drops, a flexible backup provided by energy storage, biomass and imports means that there is an almost inconsequential amount of gas use: the latter providing less than 1% of electricity in the month. Even though it is summer, there are high enough winds to cause overgeneration (times where Britain would actually have enough lower carbon power to exceed what the electricity system requires). This, of course, is electricity that could be used to support some of the additional demand from electric vehicles or charge electricity storage facilities. Indeed, the overgeneration of electricity in July actually exceeds the gas use, meaning that with more electrical energy storage to store the overgeneration July could have been carbon free.

In spring and autumn months, such as March and October (Figures 6.11 and 6.13, respectively), gas use rises in part to meet a rising electricity demand. Over these "shoulder months," there is a fall in solar generation due to more inclement weather and shorter days but a marked ramp up in wind generation due to wind-favourable weather. These months demonstrate well the value of an electricity mix; wind turbines which are less utilised in the summer start to provide more low-carbon electricity in winter and shoulder months when the solar panels reduce their contribution. Due to the large amount of wind and solar installed and their variable production, overgeneration still occurs in March and October despite the higher demand – and that overgeneration can, of course, be put to good use should there be further incentives for use of electricity surplus.

March and October demonstrate well how variable wind and solar are in real terms. It is easy to imagine wind output varying continually and rapidly over the day with each gust of wind. Similarly, it is easy to imagine that solar power output might vary continually with passing clouds shading photovoltaic panels. On a national scale, these local and minute changes in output average out meaning that the wind output generally varies over days while solar output shows relatively smooth output over the daytime. That variability is important as it means periods of sustained or near sustained electricity from renewable and nuclear power followed by periods where the flexible backup of biomass and imports are needed. Panels on your house roof might suddenly vary twenty or thirty times a day, but on a national scale that variation evens out to provide a predictable contribution to the electricity mix.

Bioenergy and imports are only so effective as backups to low-carbon sources due to their limited capacity in the vision. As a result, much more gas backup is needed during winter months when more flexible backup power and energy are needed. This higher gas use means that winter months are the most carbon intensive of the year, particularly when demand rises and wind generation falls. This can be seen occurring in the four days between 18 and 22 December when there is a sustained period of low wind and solar output and demand is particularly high (Figure 6.14). However, high winds in the rest of the month are able to provide periods of sustained low-carbon electricity generation. As a result, the interim carbon target is even met in one of the coldest and darkest months.

Questioning the vision for lower carbon electricity

The need for flexible backup from imports, biomass and gas

The way that low-carbon electricity is implemented has some key impacts on the way that different electricity generation plants operate. The role of the interconnectors which provide imports is a good example of how things behave very differently to how they did in 2017. The net import of electricity to Britain, as shown in the top of Figure 6.15, was between 0.6 and 3.8 GW in July 2017. In the vision for lower carbon electricity, due to

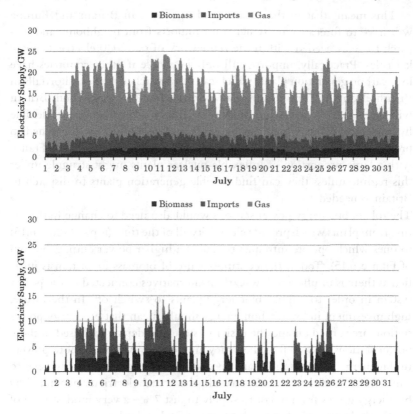

FIGURE 6.15 (Top) Net electricity imports and gas in Britain in July 2017 and (bottom) the use of imports and gas in the vision for lower carbon electricity. Decarbonisation reduces the use of imports and gas to secondary/backup supplies rather than mainstays of electricity generation. The way that the vision is simulated raises important questions around how practical and achievable this is for the way interconnectors are presently implemented and the ways that gas plants are presently designed.

the variable nature of generation mix, Britain would be producing enough electricity to remove the need for imports for several days at a time. However, since imports are used as the preferred backup to renewables over gas plants, the peak requirements for imports increase in the vision from almost 7.5 GW – i.e. it requires more interconnectors to be built and for other countries to provide flexible and low-carbon electricity generation for those interconnectors.

This means that in the future, grid operators in Britain and Europe will have to forecast requirements for imports from neighbours and dispatch these as needed within the framework of the 2030 electricity market rules. Practically, imports will only be viable if other countries have low-carbon generation plants which can be dispatched to meet shortfalls in British electricity. This means that countries like Norway, whom Britain will share an electricity connector with, might want to invest in flexible hydroelectric plants which can be easily and economically dispatched to produce electricity for other countries. However, countries like France with a large proportion of inflexible nuclear plants might struggle under this regime unless they can find flexible generation plants to dispatch to Britain as needed.

The role of bioenergy power stations would also need to change in the future from plants which provide electricity all of the time (top of Figure 6.15) to ones which operate infrequently but at a higher power output (bottom of Figure 6.15). Today, the continuous use of biomass in Britain is justified as there is insufficient low-carbon alternatives connected to the power system in order to require bioenergy plants to switch off. In the future, high investment in low-carbon energy should mean that bioenergy power stations are turned off and the easier to store and deploy fuel used in those power stations is saved for when they are needed. Although more powerful bioenergy power stations are needed in the future, these only need to run a fraction of the time. As a result, the actual electricity generated by bioenergy grows from 5% of to supply to just 7% – a very modest level of bioenergy but a critical one as backup to wind and solar.

In July 2017, Britain's gas power stations were used every day to meet the rise and fall in electricity demand. Under the vision for lower carbon electricity, these gas power stations become the last to dispatch, meaning that they are only used when there is a shortfall of energy from all other electricity plants. Reducing the volume of gas in the whole energy system, including electricity, is essential to reduce carbon emissions: 41% of electricity came from gas in 2017 but just 5% under the vision for lower carbon electricity.

Should Britain continue to invest in new gas supplies and storage?

Reducing gas use will increase the impact of and possibilities for diversifying where fuel comes from. It opens up options for replacing some of the gas power stations with dispatchable and sustainable alternatives. Such alternatives are already being discussed, studied, research, trialled and implemented. This includes using of low-carbon electricity sources to generate hydrogen from electrolysis as a fuel for export, much like gas and oil are today. Countries with lots of sunshine are consequently investigating solar photovoltaic as a means of generating cheap hydrogen and exporting this fuel around the world.

Expansion of the biogas and biomass industries, which produce fuels that are relatively easy to store, could become more focussed on the provision of backup power as these are increasingly seen as gas alternatives. Bioenergy stores share some of the key properties of gas and bioenergy plants could comprise larger storage facilities and high-power generators which are only dispatched when needed. The storage would be filled up throughout the year as feedstock becomes available, but the power station only used when variable renewables are not generating. The energy storage mix would extend beyond batteries, thermal tanks and pumped storage plants discussed in Chapter 4 and extend into the biomass available for burning.

All of these options come with various issues and controversies which need to be considered in more detail. Sourcing some forms of bioenergy can conflict heavily with the needs for food production while hydrogen is only sustainable where low-carbon shipping and hydrogen production facilities are in place.[9] Depending where the fuels come from, both bioenergy and hydrogen could also risk continuing reliance on imports which new energy seeks to minimise. However, unlike gas, both have the ability to be sourced in a responsible manner such as using waste streams to produce biogas or low-carbon generation to produce hydrogen.

Britain's ability to find a sustainable alternative to gas is heavily dependent on how much the requirement for a dispatchable filler for renewables can be reduced. The solution to finding a low-carbon means of flexible and dispatchable power to replace natural gas is ultimately likely to be a mix of different technologies including those mentioned here and others not yet commercialised.

Nuclear powered

The vision for lower carbon electricity sees more British energy coming from nuclear power stations. Nuclear is rightly a highly controversial technology with high costs, serious impacts in the event of a nuclear disaster and concerns about waste. Although climate change has to take an important role in determining our electricity future, costs and ethics must also be given high weighting. The future of nuclear power as the British electricity mix evolves to 2030 will likely be determined by the falling costs and rate of installation of low-carbon alternatives, rising demand as a result of electrification of other sectors, the economic viability of nuclear power stations to investors and the adoption of generating technologies yet to be commercialised. The increased role of nuclear power has been tempered in the vision relative to what is required to achieve a carbon intensity of electricity below 100 gCO$_2$eq./kWh and make up for shortfalls in what the variable renewables can provide. As such, the future of nuclear power should always be reassessed consummate with climate goals and how quickly other low-carbon interventions are adopted.

Variable renewables and overgeneration

Variable renewable generation sees continued growth in the vision for lower carbon electricity as a result of falling costs, adoption of new business models and improving technology. Under a principle of adopting an energy mix, the vision does not see these becoming majority providers of electricity as variable renewables sit alongside other low-carbon technologies. As a result of installing a higher capacity of wind and solar than can be used all of the time, the vision shows a number of times where more electricity is being generated by variable renewables than is being consumed. This occurs on days when it is windy, sunny and when demand is low.

It is not credible to argue that if you double wind capacity, you immediately double the contribution of wind to the electricity mix because there might not always be enough demand for that energy. For that reason, the vision for lower carbon electricity is assessed over every hour of the year. As a result, it is able to assess whether wind and solar, along with all the other electricity generators in the mix, can align often enough with demand to produce significant carbon savings.

Due to the variable output of wind and solar plants, they will rarely produce at full power. It is, therefore, prudent to design for the amount of power that can be produced most of the time to maximise the penetration of renewables in the electricity mix. An example of this is good practice design of microgrids which use solar and batteries to provide electricity in equatorial regions. Here, solar panels provide electricity during the day and charge batteries to provide power at night. A key question for a microgrid designer is determining the size of the solar array to ensure that batteries are fully charged at the end of most days. To do so, the designer will consider the amount of sun available on a typical day, rather than the sunniest day of the year when batteries can be charged by lunchtime.

On some days, it is important to accept that the microgrid will have the potential to generate more energy than is needed, as a result of having sufficient solar panels available for days of less favourable weather. This overbuilding means accepting that sometimes wind and solar plants will produce more electricity than the power system needs. It is entirely sensible to "overbuild" wind and solar capacity and on large power systems, such as the British grid, correctly specified overbuilding of wind and solar might allow a higher contribution of low-carbon energy than would otherwise be achieved.

A particular issue seen with an influx of low-carbon power, however, can be on the way that the commercial structures of the electricity grid. It has already been observed in Chile, Germany, Australia and California that when there is a large amount of solar generation prices go negative. As a result, power stations are forced to switch off and are compensated for doing so.[10] The economics of this scenario risks causing parts of the electricity system to collapse in a heap of commercial failure. It is in nobody's interest for this to happen (Figure 6.16).

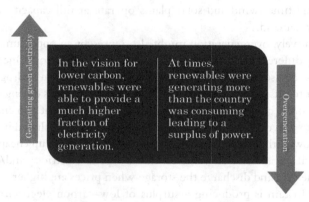

FIGURE 6.16 In order to decarbonise, there will need to be a large installed capacity of renewable power stations. At times, these will be capable of producing more electricity than can be consumed. This is not necessarily a bad thing from environmental perspectives as it means that, at other times, e.g. if there is less sun or wind, there is more electricity generation from variable renewable plants. Such overgeneration design presently works on microgrids and small island networks in producing lower cost energy. Future cost down of renewable generation or raising gas prices might also make overgeneration a good thing commercially on larger power systems.

To reduce the likelihood of these scenarios, utilities and policymakers have a number of levers they can pull and consumers can be empowered or incentivised to react to the generation of low-carbon power and consume electricity at times of high renewable supply. In these cases, the objective is to shape demand for electricity to match underlying weather patterns – something which is not implemented in the vision which conservatively assumes demand does not change. This would further reduce the carbon intensity of British electricity if it shifts demand from times when high-carbon backup plants are required to when there is lots of low-carbon power available. Such "demand response," either by using automated technologies like energy storage and smart appliances or using mechanisms like differential electricity pricing to stimulated behavioural change, should be valued for its carbon and money saving potential.

Conversely, if there is insufficient demand for electricity due to a lack of demand response, wind/solar farms would need to be switched off. Owners of renewable power stations might accept this without compensation if it occurs irregularly and is factored into their business plan. It is worth noting that overgeneration occurs infrequently in the vision as most of the time, wind and solar plants operate at full capacity to meet electricity demand.

Alternatively, overgeneration might also represent an opportunity to do something different with electricity to further decarbonisation. One possible scenario is the use of surplus electricity to make low-carbon hydrogen fuels for heat or to offset gas demand later in the year. Low prices during times of overgeneration, combined with falling storage prices, might also encourage investors to procure even more storage than is presently included in the envisioned low-carbon system. These investors would charge up cheap energy storage facilities and electric vehicles with low-price electricity and/or sell it back to the grid and discharge the storage when prices are higher.

When Britain is producing a surplus of low-carbon electricity, interconnectors to other countries might be used to sell power to neighbours. This is only possible, of course, of those European neighbours are willing to buy the energy, have a shortfall of electricity generation at home or can store the power. Doing so would reduce the need to curtail[11] power stations and would also provide income to support electricity generators in Britain. This could also have carbon benefits; if Danish wind farms are not producing enough electricity to meet their national demand at the same time that Britain has a surplus, then Denmark could procure energy from abroad rather than using fossil fuel power stations at home. Such global carbon savings are not determined in the model.

Britain could export electricity for several days under the low-carbon system, as there are times when low-carbon generators can produce more electricity than is consumed. Of course, in reality by 2030, there should be additional demand from electric vehicles which will reduce the likelihood of low-carbon generators producing more power than can be consumed and will also provide a flexible demand to better match electricity use to generation. Regardless, should there be any overgeneration or possibility of export, this is only usual if neighbouring countries have enough demand for the imported electricity and if this can be sold to them at a cheaper price than their own available power stations produce electricity.

Overgeneration is a natural consequence of the design of electricity systems with high penetration of variable renewables and has been proven to be a benefit on microgrids around the world. In Britain, that overgeneration should also be judged based on contextualised evaluation of the advantages and disadvantages.

Welcome to the store-age?

The vision for decarbonised electricity uses growth in variable low-carbon electricity sources (i.e. wind, solar and hydro) as part of a new generation mix. On some days, this results in a surplus of supply, and energy storage technologies are able to charge up using that overgeneration to store for another day. However, in the electricity mix (Figure 6.9), the contribution of storage is small in comparison to other generation types. As a result of how renewable plants have been sized relative to consumption, there are not many times when there is more low-carbon generation than demand (the situation needed for charging storage). Solar as simulated does not create enough opportunities for storage charging because it generates during the day when people are active and demand is higher, generates little in winter and is variable on a day-to-day basis. Wind also creates few opportunities for storage charging due the way the generation from turbines changes during the year. Generally, there are several days in a row where wind output is high, followed by several days of low output. This means that over the year and as modelled here, there are not many cycles to charge storage facilities from wind turbines and solar farms.

Storage provides an incredibly valuable contribution in providing services to the power grid (as outlined in Chapter 4), however, as frequently cited by those who dismiss a future energy supply from variable low-carbon renewables, massive electrical energy storage would be needed to use *all of* the output from wind turbines and solar throughout the year. The key is realising that it is not important to use all of the solar and wind electricity that could be generated. The storage simulated in the vision for lower carbon electricity gets full quickly due to its limited capacity and gets emptied quickly due to the large demands of electricity consumers. This shows that even if a low-carbon electricity system stimulates a huge growth in storage through distributed batteries and a build out of all proposed pumped hydro projects, that storage is unlikely to be large enough to take all of the overgeneration.

Storage requires a huge increase in capacity and reduction in cost to make it economically and technically viable for backing up all of the wind and solar. As identified in Figure 4.8, growth of electric vehicles (as well as increasing demand for electricity) could provide a large, flexible and alternative means of matching demand for electrical energy to production from low-carbon generation. It's my personal view that flexible charging of vehicles could relatively easily used to help align demand to supply. For example, car users could choose to charge a vehicle quickly if needed, but also have the option to put vehicles in a lower cost "flexible charging" mode to align charging rates and times to what is optimal for the power system.

In addition to building storage plants, the vision for lower carbon electricity shows that numerous mechanisms can be used to get better use of the overgeneration of variable low-carbon electricity generation. This includes, but is not limited to, switching off generators and putting in technical and economic mechanisms to encourage higher demand when renewable power output is high such as smart charging of hot water tanks. Key to determining the mix of technology used is likely to be what the market determines to be the impacts on the price of electricity.

It does not matter if this is the right vision

The vision for lower carbon power does not set out to say what should happen to decarbonise electricity, rather it tries to assess whether our present requirements can be met under a realistic and achievable 2030 electricity mix. Due to the various economic, technical and commercial constraints, predicting the electricity future is a non-deterministic problem; even if decarbonisation is a non-negotiable objective. Creating lower carbon electricity requires an ambition which respects the huge undertaking in shifting a massive, functional and life critical system onto a sustainable footing. In addition, new energy needs to come about rapidly to meet climate goals and potentially do so using technologies which are yet to be commercialised. As such, the vision proposes a methodology for evaluating new and potentially valuable technologies such as tidal as well as established technology like wind.

Regardless of the motivations for decarbonising, it is erroneous to ignore the role of economics and social factors in determining what the final electricity mix will be. Electricity needs to be generated in an environmentally responsible manner as well as being affordable. There is evidence that wind and solar electricity will reach grid parity pricing before the start of the 2020s, yet as of 2018 nuclear plants have proven to have much higher contracted prices than the wholesale market. Under the vision for a low-carbon electricity mix, over 70% of electricity comes from sources where prices can be stabilised through long-term contracts (wind, solar, nuclear, hydro, storage and potentially biomass).

The vision also aims to stimulate debate in the feasibility and utility of large reduction in carbon emissions and does fully present my personal views on how decarbonisation should happen. The assumptions and drivers that go into these models will change and so lead to different solutions. The optimal electricity mix for decarbonisation is something which is continually changing while every advance in the low-carbon toolkit affects the price, carbon impact and scalability of different electricity options. Improved efficiencies might mean that renewable technologies require less land space and are so more scalable. New manufacturing techniques might reduce embedded carbon emissions in building low-carbon generators. Behavioural change from consumers might lead to reduced electricity demand or mean that electricity is consumed at different times of the day. Societal change may demand different technologies being preferred than those considered here. Policymakers might incentivise different electricity generation or consumption patterns. Businesses might develop new models for deploying new electricity technology.

If this is true, then it is feasible that the electricity mix could contain many more wind and solar generators than in the vision for lower carbon electricity presented here. Such a scenario is considered in Figure 6.17 which shows how low-carbon future would have worked in 2017 where nuclear capacity in Britain does not change from 2017, but there is 1.5 times the wind, solar and hydroelectricity simulated in the vision for lower carbon electricity. This may be fictitious in that it might not be possible to find suitable sites for such wind, solar and hydroelectricity. However, the results are valuable in showing that a future which follows really high renewable fractions are credible from a decarbonisation perspective. In this scenario, there is actually enough overgeneration to provide a further 25% of British electricity demand with sufficient storage and/or flexible

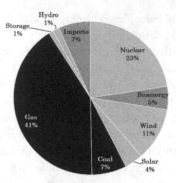

The British Electricity
Mix in 2017
280 gCO2eq./kWh

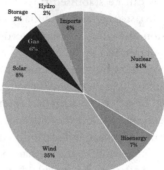

The British Electricity
Mix Under a Vision
for Lower Carbon
Electricity
55 gCO2eq./kWh

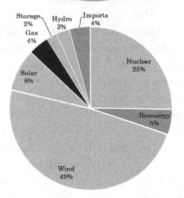

British Electricity
Under a Highly
Renewable Vision for
Lower Carbon
Electricity
45 gCO2eq./kWh

FIGURE 6.17 (Top) Electricity mix in 2017, (middle) under the vision for lower carbon electricity and (bottom) under a high variable renewable scenario with double the wind and solar generation and no new nuclear capacity. This shows how the vision for lower carbon electricity presented here is actually not necessarily the only way or best way to decarbonise. Note that the contribution of nuclear power in percentage terms has increased marginally in the bottom scenario due to a fall in overall electricity consumption.

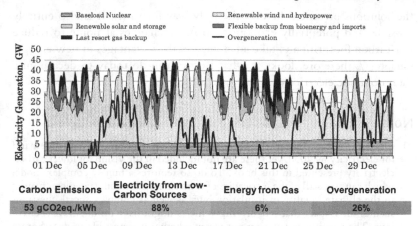

FIGURE 6.18 Electricity mix under a high variable renewable scenario with double the wind and solar generation and no new nuclear capacity in December 2017. This shows how a large amount of wind can lower carbon and meet electricity needs during the highest demand months despite inherent variability.

demand such as could be derived from electric cars or decarbonised heat. This is clearly seen in Figure 6.18 which shows the hourly electricity mix of Britain under this high renewables scenario in December 2017.

Conclusions

The vision for lower carbon electricity shows that huge decarbonisation is possible, even if the electricity mix in 2030 is different that assessed here. Before 2030 new tools in the low-carbon toolkit are likely to reach commercial viability such as tidal barrages, wave power and hydrogen power stations. In addition, the electrification of heat and transport could push up demand and could introduce much more seasonal requirements for high energy in the winter relative to the summer. It is certain that new models and methods for determining the decarbonisation path will be invented. If that is done using informed debate and a preparedness to accept new ideas, then the low-carbon transition can only be accelerated.

Achieving low-carbon grid electricity by 2030 would be a very commendable achievement for Britain. However, it is not necessary to rely on utilities, governments and corporations to decarbonise grid electricity. A variety of solutions exist, and we all have a role in developing a viable, sensible and rational future electricity mix and must not consider the solutions presented here as the only way forwards. It is also entirely possible, and potentially economic, to make changes at home to reduce our carbon footprint and take advantage of the advantages of new energy. Chapter 7, therefore, looks at what Britons can do today to decarbonise their domestic electricity.

Notes

1 Electrification of transport and heating are likely to cause increased demand for electricity by 2030, yet these are not included in the vision for lower carbon electricity presented in this book. To do so requires a hugely complex model which can assess all of the intricacies of future electrification strategies, changes to the climate, quality of insulation, etc. as well as complex social questions such as the impact that self-driving vehicles might have on our travel behaviour. The vision is not setting out to say whether it is possible to decarbonise all of British energy, it is about determining if it is possible to decarbonise what the present electricity sector supplies. By assessing the vision, it should then be possible expand this approach to assess if and how other sectors can also be transitioned to low-carbon energy.
2 For reference, the London Array in the Thames Estuary has a power of 0.63 GW.
3 For reference, there is estimated to be between 850 and 1,550 MW of remaining/unused viable hydro sites across Britain and Northern Ireland (Department of Energy and Climate Change, 2013).
4 The case for energy storage with solar is assessed further in Chapter 7.
5 Discounting effects of plant efficiencies, etc.
6 Comprising 22 GW of onshore and 25 GW of offshore turbines.
7 Ensuring there is enough electricity generation to meet demand throughout the year.
8 E.g. via new technical standards.
9 Ways of producing hydrogen including electrolysis of water (low-carbon) or processing fossil fuels (high-carbon/low sustainability).
10 In such circumstances, it is very easy to criticise utilities for paying renewables not to generate, as has been the case in the UK.
11 Deliberately reduce the output.

References

Committee on Climate Change, 2015. *Power Sector Scenarios for the Fifth Carbon Budget,* London: s.n.

Department for Business, Energy and Industrial Strategy, 2019. *National Statistics: Energy Trends: Renewables.* [Online] Available at: www.gov.uk/government/statistics/energy-trends-section-6-renewables [Accessed 30 01 2019].

Department of Energy and Climate Change, 2013. *Harnessing Hydroelectric Power.* [Online] Available at: www.gov.uk/guidance/harnessing-hydroelectric-power [Accessed 06 02 2019].

Lucas, C., 2015. *Caroline Slams the Government's Latest Energy Announcement*. [Online] Available at: www.carolinelucas.com/latest/caroline-slams-the-governments-latest-energy-announcement [Accessed 16 04 2019].

MyGridGB, 2018. *MyGridGB*. [Online] Available at: www.mygridgb.co.uk [Accessed 17 06 2018].

National Grid, 2018. *UK Future Energy Scenarios*, Warwick: National Grid.

Renewable UK, 2018. *UK Offshore Wind Capacity Set to Double Following Government Announcement*. [Online] Available at: www.renewableuk.com/news/410144/UK-Offshore-wind-capacity-set-to-double-following-Government-announcement-.htm [Accessed 30 01 2019].

Solar Trade Association, 2017. *Great British Solar Manifesto*, London: s.n.

7

A VISION FOR LOWER-CARBON DOMESTIC ELECTRICITY

There is an old adage in electricity that *"Benjamin Franklin may have discovered electricity, but it was the person who invented the meter who made the money."* That was certainly true in the old ways of doing energy however, modern technology is now providing means to beat the meter and reduce our grid electricity consumption.

While grid electricity is decarbonising, there are ways of using new energy to make our homes and businesses less carbon intensive and cheaper to run. One way of "achieving" low-carbon electricity is to sign up for suppliers who only buy sustainable energy. This influences the electricity system by encouraging investment in low-carbon energy generation; however, it is only part of a solution to reducing national carbon emissions. As an advocate of energy efficiency and as one of the first people to have solar battery storage in the UK, I have seen other means to convert your 2019 home into a low-carbon champion which is fit for 2030 and at the same time to become an active participant in that national decarbonisation. This chapter explores how much a British home can reduce carbon emissions from electricity and some of the impacts this might have utilities, consumers and manufacturers.

Energy saving and efficiency

A wise person once said that "the cheapest unit of electricity is the one you do not use." This alludes to how energy saving can have a transformative impact on personal spending power and carbon emissions. A comparison of the electricity use in an efficient home and a home with old and wasteful appliances illustrates clearly the power the efficiency can have without impacting what can be done in houses. Two homes might have same appliances and do the same amount of activities but have completely different electricity bills. One house, owned by the Smiths, for instance, might use old-fashioned equipment with an energy efficiency rating of D or below while a more-efficient home owned by the Greens uses modern versions of the same appliances with an efficiency rating of A+ and above. The Smiths and the Greens live on exactly the same schedule and behave in exactly the same way (details of the appliances used by each family is provided in the appendix). In some respects, it is like these families are living the same lives at the same pace but in parallel universes.

By only adopting high-energy appliances, the Smiths have an electricity bill of around £570 per year and their annual electricity consumption of 3,690 kWh[1] is typical for a British home with no electric heating. The Greens have a much lower electricity bill and consumption. Each year they consume just 1,080 kWh of electricity and have a bill of just £166 per year. By simply choosing energy-efficient appliances, the Greens have less than a third of the electricity bill than a family with exactly the same appliances and lifestyle (Figure 7.1).

The model to compare the families is simplistic for a number of reasons, and it does not account for the rise in consumer electronics and the numerous other gadgets in homes such as hoovers, irons, computers, computer

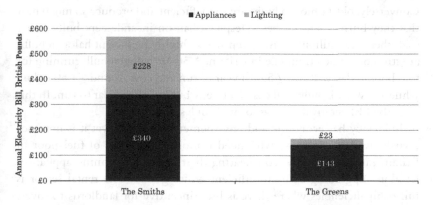

FIGURE 7.1 Annual electricity bill of the Smiths and the Greens. Electricity efficiency measures through better appliances have a huge impact of the electricity bill of the Greens regardless of their behaviour.

consoles, jet washers, steam cleaners, blenders, toasters, kettles, etc. However, the model does not need to be complex to show that reducing the energy consumption of the common appliances in our homes could save significant amounts on electricity bills with little impact on what we do. Over the past sixty years, the inefficient design of our old appliances has wasted the world's fossil fuels. Given that domestic electricity presently accounts for around a third of national electricity consumption (Department for Business, Energy & Industrial Strategy, 2018), it is clear that in the future, our buying choices will have an important influence on our ability to meet climate goals.

The fridge at the Smiths' costs £46 more per year to run than the fridge at the Greens'; however, for the Smiths to replace that fridge would cost over £300 with a simple payback of up to six and a half years. Replacing the washing machine could have a payback of up to eight years when electricity savings are considered. The economics of white goods mean that it can be a marginal investment decision to replace for more-efficient models. However, whenever families do buy a new appliance, it is always worth considering the energy efficiency as an extra £20 for a more-efficient or less-powerful model could save families hundreds of pounds over the life of the product.

The exemption to the above is lighting. The Smiths pay nearly ten times more for lighting than the Greens and the savings from switching from old-fashioned halogen and incandescent bulbs to LED bulbs are so compelling that it is arguable that most should make the switch immediately. LEDs are so good at saving energy because they produce light with very little heat. Conversely, old-fashioned bulbs are so inefficient and produce so much heat that they burn when touched! Despite a ban on incandescent bulbs being sold, there are millions in use. Even worse, highly inefficient halogen bulbs are still not banned from sale in Britain. A 50-W halogen bulb running for two hours a day will cost £5.80 a year to run and cost around £1.50 to buy while a 5-W LED bulb will cost £4 to buy but just 58p a year to run. In this case, the LED equivalent pays for itself in less than a year.

Efficiency has economic and environmental benefits, yet it is not necessarily available to those who need it most. Thousands of fuel-poor in Britain cannot afford the cost heating their homes or running appliances let alone upgrading to more-efficient white goods. The rental sector is full of inefficiencies where there is little incentive for landlords to invest in energy. The hundreds of pounds a year that energy saving can bring to the poor is money that can improve health and quality of life. It is money that makes it more like that people can afford to properly heat their homes.

Our society is weighted against the poor in many ways, and in an energy context, the fuel poor are just as important to decarbonisation as the fuel rich. Tragically, Britain already has hundreds of food banks. Should the country also have appliances refuges where those most in need can get more-efficient second-hand appliances for very little upfront cash?

Simple choices could make our homes much more energy efficient and have a large impact in reducing our carbon emissions, when they are affordable. Equally, new energy also provides ways of generating electricity in a low-carbon way and one which does not leave homes wholly exposed to the price of grid electricity. If it can be made affordable to all, solar and battery storage can have a transformative effect on domestic carbon emissions.

Solar and battery storage

All forms of generating electricity on homes will save money when they provide a cheaper form of power than the grid. Solar panels (Figure 7.2) are by far the most popular method of self-generating in Britain, and they are commonplace from the far north of Scotland to Cornwall.

FIGURE 7.2 The MyGridGB home with 3.6 kWp of rooftop solar panels and battery energy storage. Modern technology means that the existing solar array could be 50% more powerful, and batteries could have twice the capacity for less capital than was originally spent on just the solar array.

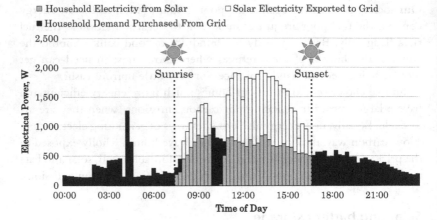

FIGURE 7.3 The basic benefit for homes with solar photovoltaic is reducing grid electricity purchases during the day using locally produced energy. Unused power is exported to the electricity grid for others to use. This chart shows the electricity mix from a British home using data, recorded during a day in May 2016.

Data from a real British home allows us to to understand the benefits of solar panels in a domestic context for a typical summer day (Figure 7.3). Overnight, the home occupier has to buy electricity from the grid to meet their demand shown in light grey because the solar panels do not provide power overnight when the sun is not shining. After sunrise, solar panels start generating electricity and by eight o'clock in the morning on the day shown, they are generating more than the house is consuming. As a result, the panels meet the domestic electricity demand and export excess electricity to the grid. At around 10.00 am, a cloud appears over the house which reduces the sunlight hitting the solar panels. Consequently, the solar generation reduces and the house is forced to buy electricity from the grid until 10.45 am when the cloud has passed and solar generation again exceeds demand. The house does not use grid electricity again until 4.45 pm when the sun starts to drop. As shown by this example, solar panels only provide energy savings during the day and more so when there is a clear sky. Very dark clouds on the day shown reduce the solar output to an unusually negligible level.

This assessment provides savings on a good summer day, but what about savings over the year? Research by Loughborough University has shown that between 15% and 69% of what is generated by domestic solar panels can be consumed in the home to reduce electricity bills in this way (Leicester, et al., 2015), and the amount saved depends on factors such as location, number of solar panels, the electricity tariff, energy demand and occupant behaviour. Some factors increase the electricity savings, for example, homes in sunnier parts of the country with south-facing roofs will generate more than homes with north-facing solar panels in the far north. Similarly, homes with higher electricity demand and occupants who are in during the day are more likely to be consuming power when the solar panels are generating electrical energy. As a result, a homeowner in the north of Britain who is in all the time and has a high electricity demand might actually save more money per year from solar panels than a home that is empty all day in the south of England with a south-facing roof, has high-efficiency appliances and a low-electricity demand.

Without some form of energy storage, solar panels can only reduce electricity use during the day. However, a battery can be charged using generation from the solar panels and subsequently deploy that energy in the home when needed. The impact that storage has on the house previously described is shown in Figure 7.4. In the morning, the battery charges up

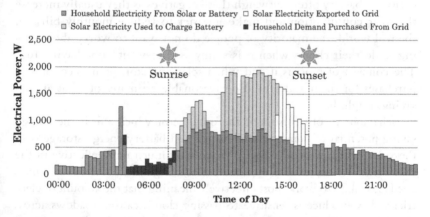

FIGURE 7.4 The basic operating principle of a domestic battery system is storing solar photovoltaic generated in the day and allowing it to be used at night. This means that the home imports less grid electricity than it would if it just had solar power. Although this means less electricity is exported to the grid, the home itself is lower carbon. As more and more renewables are integrated into the electricity grid, it could be more important than ever for homes to try and use solar power for their own consumption via storage or to charge electric vehicles to reduce demands on the grid. Present academic research is focussing on if and how to best use storage to support a grid in a highly decarbonised power system.

using solar energy and when the cloud passes over the house at 10.00 am, the battery is automatically triggered to discharge any stored energy to meet the house demand. No electricity is purchased from the grid during this time, and the battery automatically resumes charging at 10.45 am when the cloud has passed. The solar panels and battery storage subsequently meet some of the homes electricity demand overnight. In this example, the charging and discharging of the battery removes the link between when the householder is consuming electricity from the grid to when the solar panels are producing electricity: the battery makes solar electricity available when the occupant turns on their appliances. A key advantage of a battery is that it makes it much easier to reduce electricity bills from a solar power installation.

There are aspects of domestic batteries which limit their ability to save energy; they only have a finite capacity, so there is only so much shifting of solar that they can do, and 8%–12% losses in batteries means that it is marginally more efficient to consume the solar photovoltaic (PV) directly rather than via the battery. However, the advantages of owning a battery often outweigh these negatives as they usually increase the solar consumption and make it easier to reduce monthly bills. In the real world, the majority of people do not want to worry about trying to do their chores when it is sunny as they want to be having fun. The concern of most is if and when solar and storage are economical purchases for them and providing reasonable estimates of what annual savings might be.

Behaviour is, of course, not the only factor which affects the energy saving potential of domestic solar panels and battery energy storage over the year. British weather patterns make solar much more effective in the summer where there are prolonged periods of good weather; solar panels need clear skies to work most effectively meaning that the amount of electricity they produce is sensitive to passing clouds casting shadows across them. Figure 7.5 shows electricity data from a house with solar panels and battery storage over two typical summer days. The electricity demand of this house is really typical of a domestic user, with a highly variable consumption pattern as high-powered appliances like washing machines and kettles are switched on and off. During summer days, the solar panels generate much more electricity than the house consumes, and when appliances are turned on in the day or at night, the battery is discharged to support the solar and reduce use of electricity from the grid. Data from the whole summer shows that the home rarely buys electricity for months because of the solar panels and battery.

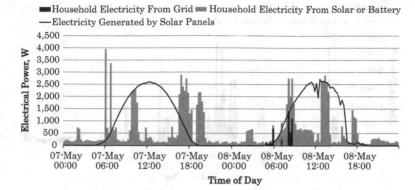

■Household Electricity From Grid ■ Household Electricity From Solar or Battery
—Electricity Generated by Solar Panels

FIGURE 7.5 The consumption of electricity of a house with solar and battery energy storage during a summer day. Grid electricity is shown in light grey, consumption directly met by solar in dark grey and electricity consumed via the battery in black. Over these two days, the only grid electricity used in the home is on the morning of 8 May. The data is taken from a home in Hampshire with sixteen solar panels on the roof and a small battery in the garage.

The winter behaviour contrasts starkly with the summer. The worst two winter days from a solar perspective are shown in Figure 7.6. Here the days are short, with only a few hours of electricity generation by the panels each day. The battery is only used for a few hours after sunset on 10 January and solar itself causes very little electricity saving. As a result, most of the chart is dark, indicating that the vast majority of the electricity used by the house during the two days shown comes from the grid.

One of the major criticisms of solar is brought by those who consider cloudy winter days like those shown in Figure 7.6. Similarly, advocates of solar power will often cite summer days like those shown in Figure 7.5 to promote the technology. Both viewpoints are very selective, and it is more balanced to assess the value of solar and storage over the whole year. Let us consider a smart home in Nottinghamshire which has sixteen solar panels on the roof solar photovoltaic system and a battery in the garage. Figure 7.7 shows the electricity mix of the home every month in 2017 where it can be seen that from April to September this solar and battery installation contributes 64% of the houses electricity. Over the whole year, solar electricity provides 50% of the total supply. There are times in the winter when the solar is ineffective, but there are also times in the summer where no electricity is purchased from the grid. In this home, the solar has

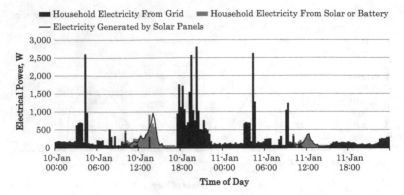

FIGURE 7.6 The consumption of electricity of a house with solar and battery energy storage during two winter days. Grid electricity is shown in light grey, solar electricity in grey and electricity consumed via the battery in black. Over these two cloudy winter days, the vast majority of electricity is purchased from the grid. The battery is empty for most of the time and so could be considered a viable and unused resource to assist the grid such as charging from the grid when national wind generation is high and/or when the power price falls. The data is taken from a home in Hampshire with sixteen solar panels on the roof and a small battery in the garage.

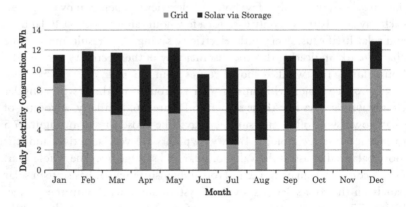

FIGURE 7.7 Monthly electricity mix at the MyGridGB smart home which has a 3.6-kWp photovoltaic system and a 6-kWh battery system (MyGridGB, 2018). Most of the domestic electricity needs are met by solar (directly or via the battery) in the summer, and in the winter, when solar production falls, most of the electricity is supplied by the grid. As a result, solar and storage are part of an electricity mix with the grid.

a significant role to play in a low-carbon domestic energy mix by working alongside grid electricity and reducing the use of fossil fuels. Solar and batteries in this case and thousands of others are effective at reducing but not replacing the use of grid electricity. As such solar, storage and a decarbonised grid can only form part of a future electricity mix for domestic properties in Britain.

This is clear in the MyGridGB smart home. The MyGridGB home was a single early adopter of solar and battery storage, and the outcomes achieved by this property can already be exceeded with more modern technology. Solar and battery technology is continually improving with products of higher quality, life expectancy and better efficiency being released onto the market. By using more of the roof space and higher efficiency panels, solar power output on the home could be increased by 75% and within twenty months of being purchased the battery capacity could have been doubled for the same money originally paid for them. If these improvements had been made, simulations show that more than 70% of the annual electricity of this home could be supplied from solar and battery storage (Figure 7.8).

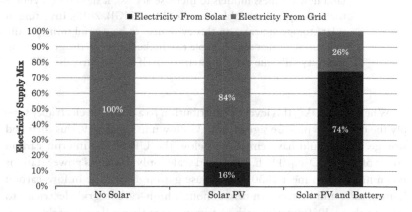

FIGURE 7.8 Share of electricity consumption from solar photovoltaic and battery storage with a 5.4-kWp solar photovoltaic system and a 14-kWh battery (MyGridGB, 2018). The larger solar photovoltaic system and battery storage than in the MyGridGB smart home means that over the year much more domestic consumption is met through electricity generated on site rather than from the grid. In this scenario, and before any electric vehicle charging, there will be a surplus of solar electricity in the summer with respect to present consumption, but that is offset by the benefits of increased generation in the winter.

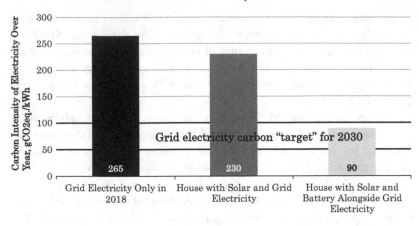

FIGURE 7.9 Carbon emissions in a home with 5.4-kWp solar photovoltaic and 14-kWh battery energy storage system in 2018. This shows that it is possible for homes in 2018, with solar and battery storage, to achieve very low-carbon emissions (even when the manufacture of panels/batteries are considered). For some but not all, this is a sensible economic choice, and with falling costs of solar/storage (and new business models to increase access), it should be a choice that more people are able to make (MyGridGB, 2018). Investing in solar/storage appears, on this evidence, to be a good thing for the environment that some people can make today. As the grid carbon intensity falls, the carbon intensity itself will in turn also fall.

When solar makes this level of contribution to a domestic electricity supply the carbon savings are significant. As shown in Figure 7.9, this solar and storage home could have emissions below the UK's 2030 interim carbon target of 100 $gCO_2eq./kWh$. On a grid scale, only using gas power stations when needed can meet 2030 greenhouse gas targets, while in low-carbon homes, solar and battery can offset enough high-carbon grid electricity to achieve those 2030 interim carbon targets more than a decade early.

Despite the potential carbon and financial savings, many people question the sustainability and robustness of solar and battery technology. Batteries and solar panels use some of the world's finite resources in the form of rare or hard to access minerals. As a result, solar and batteries can only be justifiable if they sustainably use and recycle those finite materials. To make solar and storage sustainable as well as low-carbon, it needs to be a matter of **if** and not **when**, circular economies are created for the world's batteries and photovoltaics. The solar and storage industries have started to address this; recycling of solar panels is well established through the photovoltaic cycle scheme, and battery manufacturers are starting to recognise the importance of recycling facilities. The life of components between recycling stages is also extending: batteries in the lithium family often come with a minimum of ten-year warrantees, and modern solar panels can come with performance warrantees of more than twenty-five years. Some flow batteries even are projected to last over twenty years with minimal degradation. Product performance over the duration of their operating life is rightly an important area of competition between solar and battery manufacturers and battery chemistries.

Conclusions

An investment in solar generation means that a home or business is not wholly supplied by the grid for electricity and saves money through reducing bills. Electrical energy storage such as batteries allow solar to be used at night for most of the year and remove some of the behavioural changes needed to maximise savings, such as having to use appliances during the daytime. Similarly, smart thermal storage could be used in the same way to allow solar power to be deployed to make hot water, space heat or cooling and get better energy saving from solar panels. Smart chargers also increase the amount of energy stored in an electric car battery from rooftop solar.

Solar reduces bills, carbon emissions and energy independence, and it is widely accepted that if solar and storage do not make a sound financial investment for all today, then they are very likely to before 2030 due to rising electricity prices and falling costs of panels and energy storage. Similarly, more-efficient appliances and lighting can reduce electricity bills and reliance on large power stations.

Solar and energy storage in our homes and businesses is often considered as a threat to the ways in which existing electricity systems and businesses run and the future viability of the electricity grid. To evaluate the likelihood of this happening, it is essential to remain realistic about what solar and batteries mean in the wider context of electricity decarbonisation. Not every home in Britain can install solar, particularly properties in urban areas with complex roofs or high rise flats. The price of solar and storage mean that it is presently a choice that many cannot afford or want to take. Increased loads through decarbonisation of heat and transport will also mean that homes will still need to buy some or even an increasing amount of electricity from the grid. Indeed, low-carbon homes will not just come about through solar, storage and efficient appliances alone. Houses have a carbon footprint from their construction, manufacturing of appliances and contents through to heating and transport. Solar and storage might provide a solution for reducing the use of grid electricity from early spring to late autumn, but huge leaps in solar panel conversion efficiency and battery capability are needed to generate and store enough energy to meet winter electricity, heat and transport demands.

As of the end of 2018, solar and battery technology is capable of being part of an energy mix and from a decarbonisation perspective they sit alongside other interventions in homes such as lower carbon grid electricity, heat storage, heat pumps and better insulation. The economics of solar panels to a home occupier will always be a result of social, economic and technical factors and should be regularly revaluated with falling price of panels, regulation change and revenue available from other sources such as subsidies and providing flexibility to the grid. As a result, it would be quickly become dated to present the economics of solar photovoltaic in this book. However, the impacts of solar and storage can form a key part of understanding of the impacts that new energy could have. The final chapter explores this and other concepts for the future of decarbonised electricity.

Note

1 Assuming variable electricity tariff of £0.154/kWh.

References

Department for Business, Energy & Industrial Strategy, 2018. *Energy Flow Chart 2017.* [Online] Available at: www.gov.uk/government/statistics/energy-flow-chart-2017 [Accessed 15 08 2018].

Leicester, P., Goodier, C., & Rowley, P., 2015. *Evaluating Self-consumption for Domestic Solar PV: Simulation using Highly Resolved Generation and Demand Data for Varying Occupant Archetypes,* Leeds: s.n.

MyGridGB, 2018. *MyGridGB.* [Online] Available at: www.mygridgb.co.uk [Accessed 17 06 2018].

8

CONTINUING THE DECARBONISATION OF ENERGY

"If it weren't for electricity, we'd all be watching television by candlelight," or so said comedian George Gobel in a 1954 TV show (Folkart, 1991). To ensure we have televisions, light and whatever else, we wish to power in the future means creating sustainable energy for centuries not just the next few decades. This book set out to explore and assess just part of that path to sustainability – how present electricity requirements in Britain might be provided in a decarbonised way; however, numerous challenges remain in the pursuit of transitioning Britain onto a sustainable energy system. Critically, most pathways for decarbonisation of heat and transport rely on electrification of these sectors, and this will provide new and unique challenges to the electricity system. The unanswered questions around heat and transport represent a key part of the path to sustainability. How might that be provided and what questions does it ask of utilities?

As explored in earlier chapters, decarbonisation and distribution of generation and storage within our homes and businesses fundamentally changes the economics and business cases that finance the energy system. If solar generation and batteries can allow major loads to become energy producers for most of the year, what does that mean for utilities? Will utilities exist in the decarbonised sector? Can whole towns, cities and regions ever go off-grid?

A network of millions of kilometres of cables and wires provide electricity across Britain, yet these were never designed to have hundreds of thousands of generators or to carry the high currents that might be needed to provide heat and charge our cars. Does this mean that our electricity grids will have to be built in different ways? Can Britain afford to do so?

This final chapter explores some of the key next steps facing British electricity, from provision of heat and transport through to new ways of building and financing energy. This discussion is designed to inspire the informed creative thinking that is required to continue British decarbonisation.

Plan for tomorrow, not today

One of the major lessons that I have learned transitioning from an academic researcher of energy storage to working in industry was how to correctly value the future in my models. This has meant learning to predict what the likely markets and technologies would be twelve or eighteen months from when I begin to develop business models and technical specifications for energy projects.

In the UK, a forward looking approach seemed to be taken during 2016 when developers were bidding to install batteries to provide very fast backup supplies to the National Grid. Rather than bidding using battery prices from the time of the auction, it is widely believed that one of the strategies of the winning developers was to predict or gain knowledge of future battery prices. They recognised that if they won large battery projects, it would provide the volumes and economies of scale for major manufacturers to reduce their prices or showcase their products. Whether the developers actually did this is up for debate, but it is certain that between bidding for frequency response batteries in 2016 and building them in late 2017, there was a significant drop in energy storage prices as a result of large factories opening in Asia and the USA.

Bloomberg New Energy Finance track prices across the electricity industry, providing information such as the battery indices shown in Figure 8.1. These price indices are influential and forecasts for continued falls in price of solar, wind and storage are used to justify continued investment in renewables.

It is important to understand that both technical and economic factors are causing the price of wind, solar and energy storage to fall. Wind turbines are getting larger, more efficient, increasing their output and getting cheaper to install. Solar panels are getting more efficient through new chemistries while manufacturing continues to improve and increase in scale. Battery technology is reducing in cost while learning how to extend lifetimes. For example, between the summer and autumn of 2018,

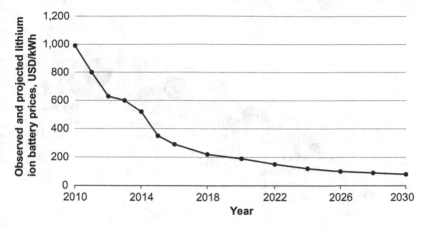

FIGURE 8.1 Lithium ion battery prices fell dramatically until 2018 as a result of increased manufacturing and technological advances. This trend is expected to continue at a decreasing rate through to 2030 and beyond. Other storage and electricity generating technologies such as flow batteries and hydrogen are expected to follow a similar trend. This cost fall gives confidence to proponents of the electrical energy storage industry (Bloomberg, 2018).

battery warrantees improved by a third as major manufacturers began to be less conservative and more willing to back the life of their products. At the same time, as technologies improved, sales volumes are increasing: the British market has grown from a single 0.2-MW battery installation in 2011 to a potential market size of 1,650 MW by 2020 (Figure 8.2). Increasing volume has given manufacturers security to build larger factories and invest in their products. Some of that initial market has come from controversial subsidies around the world, but in a post subsidy world, solar and wind are now viable investments in a significant number of international markets.

Planning for the future means designing an electricity system based on the technical and commercial realities of today as well as being mindful of the techno-economics of tomorrow. For utilities and governments, that means supporting technologies whose prices are likely to fall. It also means also putting in place the right mechanisms and investment security to ensure cost reductions for the established and emerging technologies that they want to see in the future electricity mix. For example, if a country wishes to pursue nuclear power, then they might develop policies to reduce costs, increase safety and develop local engineering expertise in the

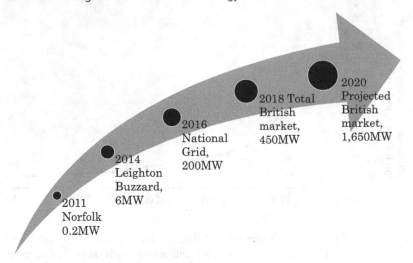

FIGURE 8.2 The market for electrical energy storage has undergone huge growth as prices fall, new markets open and new business models are found in the electricity system (analysis by author).

technology. Similarly, emerging marine technologies (tidal, wave), which have huge promise for electricity generation, might be supported through subsidies and incubator funding to develop a market so that suppliers can invest in the manufacturing and research needed to bring these products to commercial viability. This strategy has been shown to be effective with solar and wind technology. Policymaking is in part about looking at long-term electricity prices, security and carbon impacts. The mechanisms used to achieve that will differ between technologies and nations.

Having a forward looking energy programme also looks at some of the barriers in the market which discourage good behaviours. For example, adding solar photovoltaic (PV) on a new build home should costs a fraction of the cost of adding solar to an existing home. This is because putting solar on an estate of new homes achieves economies of scale to reduce costs while builders and electricians are on site being paid to do work. Would it be a bad thing, where an economic case can be shown, to make it compulsory for solar photovoltaic to be installed on all new and suitable homes in the UK? Britain mandates technical standards for energy efficiency in homes so why not do the same with self-generation technologies like solar which are cheap to install when homes are being built?

Rethinking the way that electricity is delivered to consumers

The decline in costs of solar and wind are displacing fossil-fuelled electricity plants around the world, and the energy storage industry is growing. However, one often overlooked transformation that comes with low-carbon technologies is that happening in how electricity is transported from where it is generated to where it is needed.

In 2014, I submitted my PhD which was based on three years of research into whether it was good for the electricity system for batteries to be installed in homes and besides electricity substations. At that time, storage was an infant technology with a few pilot projects being installed around the world. Although some money was also being invested in research and development to build capacity for the future energy storage industry, governments and utilities were mostly focussed on what solar and wind turbines meant for the energy system. My own research was funded by utilities companies wanting to know what residential solar might mean for their electricity networks.

Storage inspired me and my supervisor at that time for its potential to assist in the deployment of renewable energy. What the research showed was that storage did not just have a value in allowing homes to reduce their electricity bills. This started to quantify how storage could actually save the UK billions of pounds of investment by alleviating the strain on future power networks. After considering the cost of batteries, a study of just 9,163 networks in the North East of England found that storage could reduce the costs of installing solar photovoltaic by tens of millions of pounds. Storage benefits to networks have been found by many other researchers.

So why does storage work so well for power networks? The answer lies in the fact that electricity networks cost hundreds of billions of pounds to build, own and operate while upgrading or replacing a network is expensive. It is estimated by the UK energy regulator, Ofgem, that 25% of a standard electricity bill in 2018 went to network operators. Although bills are thought to be driven by the costs of fuel in power stations, it is important to remember the vast amount of money required each year to manage and expand thousands of miles of cables in the electricity grid. These networks face a major challenge to meet the additional power and energy demand of electric vehicles (EVs) and heating and could require billions of pounds of investment to facilitate decarbonisation. Increasingly, network companies are looking at energy storage to help manage this increased demand (as well as renewables) in a lower cost and more innovative way.

One of the earliest examples of using storage to help networks is a battery installed as an alternative to upgrading a substation at Leighton Buzzard in the UK. Combining battery storage for such network benefits, as well as shifting renewables, can make a strong business model as there are a variety of revenues available to fund a single energy storage installation. As an example, one UK business is using batteries to assist in the provision of rapid charging of EVs on motorways. It can be hugely expensive to upgrade the electricity grid to allow it to provide fast charging of EVs due to the short duration but high power loads that this places on cables and substations. A challenge for networks with EVs is the perceived requirement to be able to charge several cars at the same time, especially at service stations where several multiple vehicles might appear at once. Rapidly charging cars creates high power demands for twenty to thirty minutes at a time. In some places, electricity networks will fail if they tried to deliver the power needed to fast charge several vehicles at a time.

To alleviate the strain on the networks in supplying EV charging stations, stationary batteries can be installed. These allow the cars to take power from a larger battery which is slowly charged from the grid when network demand is low. In this case, the energy storage can sometimes cost orders of magnitude less than upgrading the electricity network, and one company sees this as such a strong market that they have proposed to install hundreds of millions of pounds' worth of batteries alongside new EV charging stations around Britain to protect networks and deliver low-cost vehicle energy. This protects the grid, reduces network investments and supports the role out of EVs.

Storage and distributed generation breeds new ideas that will bring about changes beyond present imagination as new technology breed new ideas. Examples of new ways of doing energy are occurring worldwide, such as in Australia where thousands of batteries in homes are being used to simulate power stations. By 2025, it is possible that you will see millions of electric cars and large batteries next to power lines to allow those cars to be charged whenever a driver wants it.

Future energy should belong to consumers as well as utilities

There is a battle in energy between those creating excitement about new ways of doing energy and those who see it as a tedious commodity that enables life. While the opening shots of this battle for new energy are being fired, millions of people and businesses still limit their engagement with energy to buying fuel at a petrol station or paying the utility bill. At the same time, some organisations are starting to create new ways for customers to engage with energy, and this is creating choices which have never been seen before by the industry or by consumers. Before the privatisation of electricity, most consumers had just a handful of electricity suppliers to choose from, yet in 2018, there are now dozens in the UK as listed on the Ofgem website. Future energy is looking to give even more consumer "choice."

In domestic or community energy schemes, consumers can now generate their own grid-alternative electricity through solar panels, wind turbines and batteries. These can be aggregated to create "virtual power plants" which replicate the size of major power stations. Batteries in houses and businesses already help National Grid by working in together as a huge and distributed backup for when large power stations fail. In the future, it is highly likely that national standards will be developed, so all new batteries and renewable power stations add to the stability of our electricity networks and so make power more reliable and stable than ever seen.

Consumerism is hugely interesting to the energy industry, particularly those promoting renewable technology. Decarbonisation via low-carbon renewables requires millions of wind turbines and solar panels to be adopted and needs business models which support mass adoption. Similar thinking is needed to persuade people to switch to EVs and electric heating and to behave in ways which support electricity and keep the cost of running networks low. For example, the energy industry is trying to find mechanisms to ensure that EVs are charging electricity is being produced. At the heart of vehicle charging is consumer behaviour and utilities need to find ways for people to want to behave in the ways that are needed to make low-carbon energy most effective. For the renewable energy industry, this means getting EVs to charge in response to the variable nature of renewable generators, i.e. to charge when it is windy, sunny or when the marine currents are flowing. For the nuclear industry, which produces power at a constant rate, this means encouraging vehicle owners to regulate charging so that the national electricity demand is more consistent.

For a decarbonising power system, consumerism risks being ineffective if it needs total participation to be successful. People have different personal objectives and will often make decisions which might seem illogical. For example, despite the revolution in smart phones, your nana might always buy an old mobile phone! Not everybody has the will or time to engage, and for some, energy will always be something limited to a bill with their supplier. For example, a corner shop might make very good money installing solar panels on their store; however, if energy makes up a small fraction of costs, then it is difficult to justify any time or effort assessing solar relative to investments in other parts of the business.

To enable the low-carbon energy transition, there will need to be hundreds of different business models to appeal to different customers and for different technologies to be successful. On one end of the scale are the large investments needed for nuclear power stations, interconnectors and offshore wind farms. On the other end of the scale, businesses will need to be set up to open access to new energy to all – including those who and who cannot afford capital investment for things like solar panels. For some, that might be low-carbon energy supply contract. For others, it might be community energy generation or storage schemes. For vehicle manufacturers, it might not just be developing the technology of EVs but also opening up new ownership models to make them affordable and available to all. Some companies will need to be set up to ensure greater access to affordable energy and so reduce the fuel poverty which kills hundreds every year in Britain. New energy is in part about resolving decarbonisation but could also improve the value that the power system brings to the population and business.

That is as true for consumers as it is for the companies which run electricity systems. Historically, electricity networks were set up as the only way to transport electricity from large power stations to our homes and businesses. Since the dawn of mass adoption of electricity it is the networks that have had a monopoly on the transportation of power, and within that, it is the technical and regulatory standards that underpin how engineers and energy companies behave. That monopoly was justifiable with the old energy of large power stations and millions of consumers, yet there is increasing evidence that model will have to change to enable the full potential of distributed energy to be realised.

As monopolies in the transportation of power, it can be argued that network companies have a responsibility and incentive to allow creativity and innovation. In the future, the electricity system will have to facilitate consumer choice. Their role should be allowing engagement with power in thousands of different ways such as aggregating small power stations for some, old style utility contracts for others, community energy schemes, large power stations, fixed term contracts, etc.

To grow their businesses, network companies should be looking at new energy as an opportunity to open new revenue streams and to be the key stone to the energy transition. For example, a community energy scheme under the new electricity paradigm might be a village that wants to install a wind turbine and a battery. The village would have historically purchased electricity from large power stations via an incoming supply.

If a small village wishes to set up a consumer energy scheme, they might blend the electricity generated by the wind turbine to power their homes with electricity from major power stations using the existing electricity grid. To do so, they need to connect the wind turbine to the electricity network, send the power over the cables in the village and deliver it to where it is needed in homes. That could be a new revenue stream for the company who owns the network and at the same time help enable growth of low-carbon generators.

The community might also install a battery to provide backup power if the main supply to the village breaks. The incoming supply both to and within the village is owned by the network company and so they are culpable for the main supply failing. If the incoming supply fails, customers in the village could transport power from the battery to their houses using their local network to keep the lights on. The network company should allow this so that their customers are less likely to notice any issues with the electricity supply. For both the battery and the wind turbine in this example, it is the network company that holds the key to letting new technology work for consumers. They are key to the transition and must not put technical or commercial barriers in the way of adoption of new energy (Figure 8.3).

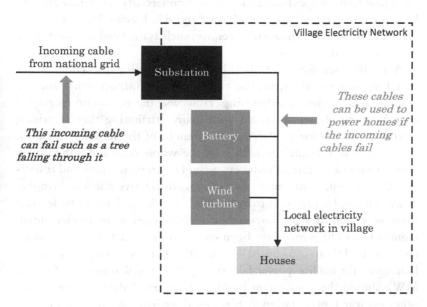

FIGURE 8.3   A battery connected to a local electricity network can be used to provide power if the main incoming supply fails and only if the owner of the network allows it to happen.

Learn from the successes and failures of subsidies to support new technologies

As demonstrated in Chapter 1, the fossil fuel requirements of other sectors will also need to be decarbonised and lower carbon electricity is only part of the transition needed in British energy. However, it is widely reported that there is not enough potential for wind and solar power alone to decarbonise the whole energy system beyond electricity, which is one conclusion that is drawn from David MacKay's seminal book "Sustainable Energy Without the Hot Air."

There are hundreds of ideas for ways of storing and generating low-carbon electricity and heat. Well-known technologies such as geothermal power, tidal stream generators, wave power, nuclear fusion and tidal barrages are just some of the underdeveloped low-carbon means of producing electricity and/or heat in Britain. Similarly, hydrogen and other forms of energy storage have theoretically promising advantages which are yet to be exploited. Although high costs, a lack of practical deployment experience and/or ongoing research requirements limit the commercial viability of these technologies at the time of writing, the wind and solar industries have taught us valuable lessons, both good and bad, in how to transition these new technologies from concepts to commercially viable investments. It is worth remembering that both solar and wind technologies were in a similar position before financial incentives such as the Feed-in-Tariff (FiT) were established.

A justification for these subsidies was socialising the costs of not switching to renewables, i.e. the costs to health, infrastructure and the environment by not decarbonising. However, the economic incentives have been hugely controversial with many attributing them to rising electricity costs. One rarely contested benefit of the subsidies has been the creation of revenue confidence for the whole solar and wind supply chain from raw material producers, manufacturers, retailers and installers. This revenue confidence has had a transformative effect in bringing down wind/solar prices. It is irrefutable that lessons have to be learned from how subsidies for low-carbon technologies were implemented. Contracts for Difference have been successful in reducing offshore wind costs in the UK to £57.50/MWh in 2017. However, the Contract for Difference for nuclear power falls from £92.50/MWh to only £89.50/MWh should the consortium building Hinkley C decide to build another nuclear power station at Sizewell (Department for Business, Energy & Industrial Strategy, 2016). Subsidies for solar power were not set with a consistent and predictable price reduction, and the solar industry was so successful at leveraging the subsidy that install rates would rise to extremely high values. As a result, the government intervened with a series of sharp subsidy drops to temporarily make solar uneconomic in the UK. This costs thousands of jobs while a climate of fear created by a rush to complete installations before subsidy changes also fostered poor workmanship by inexperienced builders which has damaged the reputation of the solar industry. A more managed and steady decline in subsidy for solar would probably have led to a more controlled install rate which supported jobs.

That being said, subsidies worked to bring down costs because they were effectively bankable energy contracts to justify investments in high capital energy technologies like wind, nuclear power, solar and batteries. National Grid learned the lesson so well that, to a certain degree, they mimicked it in 2016 when they launched their procurement exercise for fast-responding battery storage. They launched a competitive auction where different players could bid products in at the lowest price to meet a prescribed technical specification. This led to some of the cheapest energy storage installations the world had seen at that time. In effect, National Grid mimicked the key part of the subsidy which was long term and bankable contracts and let the market work out the cheapest way to provide a service.

Subsidies and recent power auctions have taught policymakers many good lessons, and it is worth considering whether subsidies to support other energy technology should now be implemented. In the UK, it is now time to consider how to support technologies with potential to expand the low-carbon energy mix. This includes technologies for low-carbon heat (hydrogen, geothermal, etc.) and underutilised forms of low-carbon electricity generation (tidal stream, tidal barrage, etc.).

Britain is addicted to gas and needs to diversify away from it

In 2016, Britain saw the first hour of coal-free electricity, yet there has not been a single hour of gas-free electricity this millennium as gas power stations run twenty-four hours a day. Gas was responsible for nearly three quarters of the carbon emissions in British electricity in 2018, and when there is a shortfall of electricity from renewable power stations, the country usually switches on even more gas power stations to ensure demand is met.

Gas is an easy-to-store, highly energy dense and dispatchable fossil fuel which Britain has transitioned to dominating too much of the energy mix. In 2010, 66% of total energy in UK homes was attributed to gas for space and water heating needs (Hamilton et al., 2013). Not only is gas critical to the economy and our comfort, the use of this fossil fuel is highly seasonal as during the winter, gas use can peak at over five times what it would be in the middle of summer. Nuclear power plants which economically run twenty-four hours a day and low-carbon generators are not presently designed to completely alleviate seasonal gas demand. To further sustainability, there is a huge need for efficiencies and low-carbon flexibility in the heat and power sectors (Figure 8.4).

How Britain get its gas is also changing, and since 2011, the country has been a net importer of the fossil fuel. In 2001, 98% of British gas came from local production which was considered to be a more secure source than relying on foreign countries for a stable and affordable supply. That changed as North Sea production fell, and many are now questioning the security of British gas supplies, the ethics of where it comes from and whether the country might one day be held to ransom by foreign powers putting restrictions on supply.

What would happen to the economy if there is a shortfall of gas? A disruption to gas supplies would have dramatic impacts on the electricity system as Britain does not have sufficient numbers of coal power stations to switch to if there is a gas shortage. Without remedial policy decisions, there could be power outages or rationing of electricity unless reserves in gas storage facilities are sufficient in the event of supply disruptions. A gas shortage

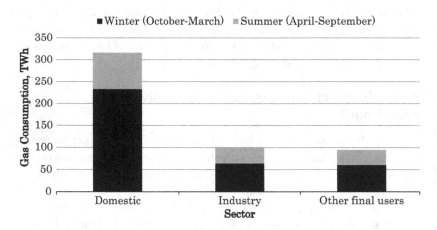

FIGURE 8.4 Gas consumption in the UK. Gas is the primary source of space heating in homes and decarbonising that winter heating requirement is a key challenge as Britain tries to break its addiction to gas (Department of Business, Energy and Industrial Strategy, 2017).

in the winter of this type could also lead to millions being unable to heat their homes with potentially fatal consequences for those most vulnerable and economically disadvantaged in our society. Price rises as a result of gas shortages rise would increase fuel poverty and could lead to a collapse of established energy suppliers. Unless there is diversification of the energy economy away from gas, or improved security of supply, the consequences of a loss of British gas could be extremely serious (Figure 8.5).

The drive to open up shale gas extraction in the UK is in part a reaction to try to secure local supplies. Shale gas extraction is, of course, short sighted given that those local gas resources will one day run out and due to the dangers it imposes on local environments Extracting and burning shale gas is using up finite resources which can never again be used either for fuelling energy industries or as a feedstock to industry.

Diversifying the electricity generating industry through the low-carbon toolkit is a much more sustainable way to reduce dependence on foreign gas while decarbonisation can occur if gas is viewed a backup to renewables when they are generating insufficient electricity to meet demand. However, in 2018, gas was part of the electricity generation mix every hour of the year because there had not yet been sufficient investment in low-carbon alternatives. With present rates of growth in low-carbon electricity, it is likely that Britain will see a gas-free hour of electricity generation before the end of the 2020s, particularly if there are strong winds over Christmas or New Year when demand is usually low[1]

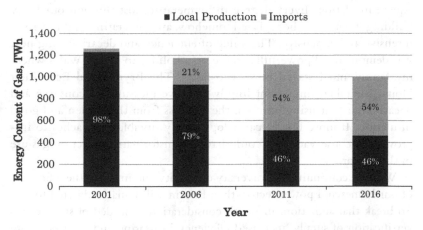

FIGURE 8.5 UK gas imports and production (2001–2016) showing how and when the UK changed from a net gas producer to a net gas importer (Department of Business, Energy and Industrial Strategy, 2017).

or if sunny weather coincides with windy weather on a summer weekend when demand is also low. That will be one of the most seminal moments in Britain and could represent the first low-carbon electricity generation the country has ever seen.

In addition to reducing gas use in electricity, over 80% of British homes need to switch from gas to alternative heating sources to bring about decarbonisation of the whole energy mix. There are many ways in which to achieve a decarbonised heating supply and to break the gas addiction. If the switch is from boilers to electric heat pumps, then there will consequently be increased demand for electricity which will need to be met. It would be a tragedy from a sustainability perspective if that increased demand was only met by gas power stations. There is a need for a policy focus and research into the role of seasonal and inter-day storage for matching heat production to demand and alternative heat technologies like geothermal power. Technologies such as phase change materials for domestic hot water storage might be easily and economically scaled to provide high efficiency, low loss thermal tanks. This might enable a week of windy weather to generate and store heat, which is then used in a subsequent cold spell to warm homes. It has been reported that thermal storage is not getting the policy focus that it might deserves (European Association for Storage of Energy, 2017) and that is probably true.

Reducing gas use is not just limited to finding alternative ways of generating electricity and heat. Our homes and businesses leak heat out to the environment through gaps in our windows and doors and inadequate insulation. Every degree in temperature lost through our leaky building stock has to be replaced somehow, and generating heat is carbon intensive and expensive. The value of efficiency and decarbonising heat was demonstrated powerfully to me on a holiday to North Wales where my family and I stayed in an eco-hotel. The Bryn Elltyd Eco Guest House has taken simple but highly effective measures to conserve and decarbonise heat using porches, the biomass from the garden and good insulation. If more houses can adopt locally suitable approaches to improve heat conservation, it could have a transformative effect on gas use in this sector.

Without continuing the diversity electricity and improve the efficiency of our homes and power sector, then Britain will remain addicted to gas. To break that addiction, in-depth consideration is needed of storage, diversification of supply, increased efficiency in heating and reduced waste of heat itself. Failure to do so risks severe consequences to energy poverty, safety and our economic security (Figure 8.6).

FIGURE 8.6 This eco-hotel in Wales is recognised as a leading building for showcasing innovation in insulation and low-carbon heating. The hotel also does a great breakfast (image by author).

New ways of paying for electricity

Shifting the power system from operating a small number of large power stations to millions of distributed generators and batteries will cause significant changes to the way it is funded. As shown in Chapter 7, a home with solar and battery storage might be nearly completely off-grid for most of the summer when there is favourable weather but import a significant amount of electricity from the grid in the winter when solar panels produce much less electricity. If there is mass adoption of solar and battery storage by households and businesses, electricity suppliers and generators will need to learn how to finance an energy system with prolonged periods of selling much more electricity to homes and businesses in the winter when the solar generation is less. That challenges a business model which meets the majority of electricity demands all year round. In addition to consuming electricity, a home could become a more active participant in electricity. Solar power which is not used in the home in the summer could be exported for businesses and homes who cannot install solar. Similarly, spare capacity in batteries could be used to help manage the power grid such as charging when there is an abundance of generation from offshore wind turbines. Numerous companies called aggregators are looking at ways of utilising that underutilised battery energy storage capacity to provide a variety of services such as frequency support, carbon trading and overnight battery charging. Financial incentives can be also used to encourage behaviour or technology adoption, such as encouraging battery charging to raise demand when it is windy or overnight to support nuclear power stations.

Some researchers also think that how the energy system is financed and regulated might change, for example, consumers with solar and batteries might be given cheaper power if the solar and battery can be used to support the wider electricity grid and make supply more stable. One highly controversial idea is that, rather than being billed for every unit of electricity that is consumed, consumers will pay a monthly "subscription" fee to be connected to the electricity grid, much like the way phones and cable television are paid for. This is seen as a way of providing revenue security to electricity network operators, who rely on year-round electricity consumption to fund their cables and transformers. Paying a subscription fee would provide a continuous revenue for the electricity industry but would need to be enacted carefully to keep electricity affordable and to continue decarbonisation. Presently, electricity is billed relative to consumption (the more you consume the more you pay) which provides a financial incentive for energy efficiency. Important moral questions need to be asked around the way energy is billed in the future if it is to continue to encourage efficiency while at the same time it is providing the income needed to invest in and manage electricity networks.

Although not modelled in this study, electrification of transport and heat could have two significant effects on the way that energy and transport systems are paid for. First, the additional demand from EVs will need to be met through new-generation plants. There is also an uncertain cost expanding electricity networks to allow energy from power stations to be delivered to charging points. Smart thinking can go some way to reducing the costs of these network upgrades, as any investment in the networks is ultimately paid for through electricity bills. Although some of these costs will be met through the additional revenue from purchases for the electricity to vehicles and electric heat, there is a risk that these costs could also raise the underlying costs of electricity. There is also a significant tax on existing transport fuels such as petrol and diesel. If vehicles are electrified, then the tax shortfall could be made up through increased tax on electricity, which would need to be carefully implemented so as not to increase fuel poverty for those without their own vehicles.

The issue of poverty is also hugely important in the future electricity system, in the issue of increasing access to those least able to pay. Domestic solar photovoltaic is a great opportunity for the poorest to reduce their bills, yet the upfront costs required are by their very nature unaffordable to those who are energy poor. Mass adoption of solar photovoltaic on roofs is one route to decarbonisation of electricity but requiring homeowners to find upfront capital to do so will always be a barrier to access. For those in energy poverty, external financing mechanisms will be required for access such as through social housing providers. Given the value of solar in the energy transition and reducing electricity bills in the future, that is something that needs to be encouraged.

The role for regulation and government in changing commercial realities

Governments, policymakers and electricity industry regulators influence factors from the provision of safe and reliable power, to keeping markets competitive and ensuring value for money for customers. In a decarbonised electricity system, with changing types of power station, there are strong arguments that markets and regulators need to adapt to function effectively in the future. The following brief examples indicate why this is the case.

Most low-carbon technologies (including nuclear, wind, solar and hydroelectricity) have a very similar investment profile with high upfront cost and low operating costs. As such, investors are required to spend huge sums at the start of the life of a low-carbon power station and recoup that money through selling electricity over its life. If there is little or no security on the price that the investor will be paid for electricity, then the power station will be perceived as higher risk and will demand a higher return. The fact that investors will be seeking more money has particular consequences such as a marked slowdown in the rate that low-carbon generators will be built and higher prices for low-carbon electricity. To reduce the costs of low-carbon electricity to consumers, regulators need to find ways of stimulating competition, valuing quality engineering and ensuring investors in low-carbon energy to have access to low-risk contracts. As the cost of wind and solar fall, the cost of finance and expected return actually begins to dominate as a factor which determines the final electricity price. Policymakers and utilities need to recognise that fact.

The rate at which low-carbon investments are made is important and will be the result of factors with numerous dimensions. A faster build out meets climate goals faster, yet it is widely predicted that over the 2020s, the price of all forms of renewable power will continue to fall as technology and manufacturing improves. So should Britain build low-carbon generators today if prices will be lower tomorrow? To keep prices of most forms of low-carbon power falling, it is essential to maintain a strong international and local market so that the required research and development of technology and manufacturing continues. Therefore, investment today is essential to ensure low prices in the future. That might also mean building higher cost low-carbon generators today, e.g. nuclear, if these are shown to have low prices in the future when compared to the costs of alternatives (including climate change impacts, grid integration and storage costs).

Another key role for policymakers is ensuring that the skills needed to run the evolving electricity system are maintained in the face of decarbonisation. Changing electricity systems might appear to be a threat to the jobs of people who already work in the sector, while the nature of solar and wind farms mean that they create lots of jobs during construction but very few during their operational life. It can be credibly argued that some of the cost of nuclear power in Britain actually relates to the fact that there is little recent indigenous experience in building nuclear power plants. No nuclear power stations have yet to be built in Britain in the 21st century and closing out the build programme in the 1990s meant a partial drain of skills and expertise. To commit to a low-cost and low-carbon electricity mix, policymakers need to provide a continuum of stability to maintain jobs and skills in the country.

The role of governments, policymakers and regulators is critical to achieve decarbonisation. The above are just some examples, and as Britain decarbonises, it will be critical to ensure that technologic and commercial advances are matched in the way electricity is governed. That can be achieved if there is strong and informed debate about our decarbonised new energy future.

Rethinking what storage can do for electricity decarbonisation

It is probable that electrical energy storage in British electricity in 2030 will differ to how it was in 2018. There will be a massive decline in the storage in the form of coal heaps, yet there should be an increase prevalence of batteries and other forms of energy storage in homes, businesses, transport sector and the electricity grid as costs fall. If the economics are right, there could be a build out of new pumped storage plants around the country. Wind and solar farms investors are already assessing storage plants to sell their power at peak periods rather than whenever the sun shines or wind blows. Similarly, a proliferation of EVs is expected to both grow the demands for electricity as well as increase the amount of flexible electricity storage available to match demand to power supply. This could be supplemented by flexible stores of heat for our hot water, and potentially space heat, to further decouple supply and demand. Welcome to the store-age!

In the right scenario, storage can be part of a transformative mix of technologies to bring about decarbonisation such as in a domestic context, yet it is important to recognise that is not always the case. The vision for low-carbon electricity identified a need for more *long-term storage* which can help the power system mean peak demands in the winter when gas use is highest using electricity generated weeks or months before. Seasonal storage undergoes just a handful of cycles per year, unlike domestic batteries which undergo hundreds of daily cycles to facilitate charging and discharging from solar panels. At present, some of Britain's seasonal storage comes through gas and coal bunkers. Identifying and supporting commercially viable, seasonal and fossil fuel-free electrical energy storage is an important challenge for the decarbonising power system. The scale of the requirements for this storage increase further if heat were to be electrified, as this would increase the demands for electricity in the winter more than in the summer.

In addition to shifting low-carbon generation to when it is needed, storage has a wide variety of roles such as maintaining short-term imbalances in power, backing up power stations, helping manage networks and reducing customer bills. All of these applications require an energy storage mix. This is a suite of evolving technologies which provide various services to the electricity system where there is a need and when this has a demonstrable net benefit. If storage technology improves and high capacity seasonal storage is shown to work economically and technically at scale, then expect to see deployment of mass storage to shift solar from summer to winter and wind from blustery to still days. If storage becomes affordable to all, expect to see it in most of our homes helping to keep our bills low or moving solar generation from day to night. If storage becomes more affordable to networks, expect to see batteries at your local substation. If storage becomes more affordable and high density in a transport context, expect a rapid uptake of EVs.

Given the rate of decline of costs in the storage industry, and the variety of business models being explored by industry, it is highly likely that the storage industry will only grow as the decarbonisation of electricity continues. However, in a changing world, it will only take one small shift in technology, a flexible low-carbon electricity generating technology, a switch to hydrogen vehicles or something as drastic as demand itself that can super-flexibly respond to generation for the future of storage to change.

Decarbonising heat and transport

The major headwind against electrical decarbonisation of electricity is the huge burden as a result of transitioning transport from petroleum and diesel and heat from natural gas over onto the electricity system. As identified in Chapter 1, energy resources consumed by transportation and heat is even more significant than those from the electricity sector. Decarbonising these could create both higher electricity demand and bigger differences in consumption between seasons. Quantifying what this means for electricity is beyond the scope of this book, but that should not downplay the importance of the issue. That being said, there are some important factors which need to be considered in assessing this issue.

First, transport and heat are particularly inefficient users of the fuels that they consume. The rate of heat loss within British buildings is huge when compared with what can be achieved in some of the most efficient buildings around the world. EVs are much more efficient in their use of energy than the internal combustion engine (the latter of which produces large amounts of useless heat alongside the power to move vehicles). Demand reduction through efficiency has been shown to have major importance in decarbonising Britain's existing electricity system. The impact efficiency can have on transport and heat is likely to be even more significant due to the volumes of energy that can be saved.

Second, the seasonal impact of heat, i.e. a high winter demand, should not be ignored. The technologies which are used to make heat will need to be economically and technically viable even if only used for half the year – i.e. in winter not summer. It is also important to look at how existing infrastructure can be adapted and trials of injecting hydrogen and bioenergy into existing gas networks might have a short-term impact in reducing carbon emissions and diversifying supply. Decarbonised heat does not need to come entirely from low-carbon electricity, and it is worth spending some of your time researching alternative heat supply technologies such as geothermal (heat from the earth) and heat from waste. Total energy requirements for heat can also be reduced using measures such as heat in abandoned mines to improve the effectiveness of heat pumps and, of course, the huge energy savings as a result of vastly improved efficiency and insulation in buildings. Heat requires a mix of solutions just as much as the electricity supply evaluated in this book does.

Lastly, heat and transport can have some elements of flexibility which is useful in a low-carbon electricity system. EV batteries (and potentially hydrogen fuels) are potentially convenient ways of decoupling between when electricity is produced and when it is consumed, and this flexibility is important for supporting all forms of low-carbon electricity generation (nuclear, wind, tidal, etc.). Heat is relatively easy to store, so the energy storage mix presented in Chapter 4 might also need to include more heat as well as electricity storage.

Decarbonising heat and transport is a challenge, but the methods examined in this book, if not the exact way decarbonisation is implemented, will be useful in determining how to achieve a multi-vector low-carbon energy future.

Conclusions

Although it's clear that decarbonisation is likely to be a good thing for society by reducing the likelihood of devastating climate change, it would be good if other tangible benefits also result from the new electricity system. I was once asked what the biggest barrier to decarbonisation is and the key word to me in our electricity future is "ambition." Ambition means backing new technologies and giving certainty to society and to industry. Failure to back nuclear power in the 1990s severely damaged Britain's ability to build new plants in the 2020s. Only by rebuilding and supporting the nuclear industry can ensure it can economically compete with fossil fuel plants. Policymakers must not make the same mistake with solar, wind, marine, hydro and the energy storage industry as these technologies emerge into commercial viability. The key difference with new energy is that the public doesn't have to wait for utilities to build new power stations or be a choice less victim of higher prices or falling quality of service. New energy means an ability to engage in electricity like never before, and I hope this book sparks your imagination in how you might change your behaviour and investments to bring about better energy for you.

This book set out to explore how present electricity requirements in Britain might be provided in a decarbonised way. Whether you believe the motivations for renewable, low-carbon energy or not, it is undeniable that their role is increasing. The journey to full decarbonisation is, of course, incomplete; however, I do hope that I have shown that such a future is possible. The numbers presented on volumes of low-carbon energy are entirely achievable, and the vision for lower carbon electricity mirrors what trade bodies of the various low-carbon industries such as The Nuclear Industry Association, The Solar Trade Association and Renewables UK have set as targets or ambitions for their industries. I've only relied on widely deployed electricity generating technologies and have omitted emerging technologies such as tidal and wave power which could, of course, add to our low-carbon future. With credible evidence, I would recommend adding these generators into the low-carbon toolkit and so assess what they can add to further electricity decarbonisation in Britain and other countries around the world. Ultimately, I hope to have shown that Britain can install enough low-carbon generation to achieve decarbonisation of present electricity demand. The question that should be asked now is not if decarbonisation of electricity is possible, it is when that change can be made and whether we are ambitious enough as a society to make the necessary changes.

Note

1 Demand falls at Christmas as a result of a lack of industrial/commercial activity.

References

Bloomberg, 2018. *Coal Is Being Squeezed Out of Power by Cheap Renewables.* [Online] Available at: www.bloomberg.com/news/articles/2018-06-19/coal-is-being-squeezed-out-of-power-industry-by-cheap-renewables [Accessed 20 01 2019].

Department for Business, Energy & Industrial Strategy, 2016. *Hinkley Point C.* [Online] Available at: www.gov.uk/government/collections/hinkley-point-c [Accessed 30 01 2019].

Department of Business, Energy and Industrial Strategy, 2017. *Energy Trends 2018: Table 4.1 Natural Gas Supply and Consumption*, London: s.n.

European Association for Storage of Energy, 2017. *Thermal Storage Position Paper*, Brussels: s.n.

Folkart, B. A., 1991. *TV Comedian George Gobel Dies at 71.* [Online] Available at: www.latimes.com/archives/la-xpm-1991-02-25-mn-1417-story.html [Accessed 16 04 2019].

Hamilton, I., Steadman, P. J., Bruhns, H., Summerfield, A.J., Lowe, R., 2013. Energy Efficiency in the British Housing Stock: Energy Demand and the Homes Energy Efficiency Database. *Energy Policy*, 60(September 2013), pp. 462–480.

EPILOGUE

By the end of 2018, decarbonisation efforts in Britain had resulted in a rapid reduction in coal use, falling electricity demand and growth in low-carbon renewable generators. Although these changes led to the greenest electricity for several generations, it can be argued that Britain's energy system was in its most perilous state for half a century. Decarbonisation efforts have left the nation dependent on gas for over 40% of electricity and nearly all winter heat requirements. As a result, any disruption to gas supplies would have disproportionate effects on our lives and our economy. Going forwards, the country has to continue to diversify energy supplies in order to continue the decarbonisation journey and break the resilience on gas. In this book, I have presented a vision which shows that decarbonised electricity can be achieved although it is somewhat irrelevant whether decarbonisation is achieved exactly as modelled in this book. There are thousands of ways of reducing our greenhouse gas emissions using the continually evolving low-carbon toolkit presented in Chapter 3, and it is important that decarbonisation continues in an informed and considered way. Through reflecting on the things you agree with and disagree with in this text, you should now be more informed on the role that you might play in that journey.

The inconvenient truth in all decarbonisation strategies is the importance of people power. That is to say, that if humans are required to widely adopt the renewable technology, higher efficiency and lower carbon behaviours that are needed to decarbonise electricity, then it cannot be the acts of a handful of women or men that lead our transition. As a result of the mass change that is needed, it is impossible for the few to micromanage the job of the many. As a reader, you are a self-selecting audience member who will have an inherent interest in the subject. To act upon your ideas requires you to always recognise that they will be doomed to fail unless they can and will be delivered by millions. Achieve that, whether through buying new light bulbs, inventing or investing in low-carbon technology or decarbonising your businesses, and you become part of a collective human effort to overcome runaway climate change. As a consequence of being part of change by millions, you are carrying out actions that improve the planet for all.

APPENDIX

Calculating carbon emissions: example

The vision for lower carbon electricity focussed on whether Britain's electrical demands can be met in a lower carbon manner. For each scenario assessed in this book, carbon emissions are estimated using mean carbon emissions factors from the Intergovernmental Panel on Climate Change (IPCC; Schlömer, et al., 2014). To assess the carbon emissions for a generation mix, the total electricity from each electricity generation type is summed and then multiplied by the carbon intensity to give the total carbon emissions as shown in Table A.1. Dividing the total emissions by the total electricity production gives the carbon intensity.

Key weaknesses in the electricity system modelling

- An assessment of the ability of the electricity grid to integrate low-carbon electricity generation is not modelled. This is deliberate as the model seeks only to assess whether there is enough energy available to lower carbon emissions, rather than determine how power system engineers might make that happen. Managing a grid with a high volume of variable renewable electricity generation is complex but a solvable issue.
- Life-cycle greenhouse gas emissions of energy sources used in this study are the median values from the IPCC as identified in Figure 1.5. There are a number of other approaches that could be used, such as marginal

TABLE A.1 An example of the calculation of the carbon intensity of electricity

Generation type	Total electricity generation, GWh	Median lifetime carbon intensity, $gCO_2eq./kWh$	Calculated carbon emissions, MT
Nuclear	1,102	12	13.2
Biomass	225	230	51.8
Imports of electricity	393	12[a]	4.7
Large hydro	24	24	0.6
Wind	239	11	2.6
Solar	249	41	10.2
Gas	1,990	490	975.1
Coal	185	820	151.7
Total	4,407	n/a	1,210
Carbon intensity	275 $gCO_2eq./kWh$		

a   Notional value for calculation example.

carbon emissions which reflect the additional carbon for every unit of energy. However, with such negative press surrounding lifetime carbon in low-carbon technology it felt important to assess on these terms.

- Data for the decarbonisation models is taken from Leo Smith's Gridwatch and from BM Reports which likely underestimates the capacity factors of wind and solar farms (this is because it is difficult to know exactly how much wind and solar capacity is being measured at any one time). As a result, scaled capacity factors are used in the analysis in Chapter 6. Wind capacity factors are modelled at 30% which represents a blend of offshore and onshore technology. A shift to offshore technology is likely to increase capacity factors. Solar capacity factors are modelled at 9.7%.
- The model should be expanded to cover multiple years of data rather than just a single year to study the impact of varying annual weather patterns.
- Wind and solar generation are scaled according to the increased capacity in the vision for lower carbon electricity. This means that aspects such as the increased wind speeds for offshore wind turbines (and consequent higher capacity factor) are not modelled.
- Requirements for the stability of the power system are not modelled, such as the requirement to balance instantaneous demands for electricity on a second-by-second level. As discussed in Chapter 4, some of this role can be filled by flexible technology like electrical energy storage.

Electricity use by the Smiths and the Greens

Table A.2 summarises the electrical equipment used in the Smiths' and the Greens' houses. The Smiths use old-fashioned, high-energy appliances, while the Greens use high-efficiency equivalent devices.

TABLE A.2 Comparison of the electrical equipment in the Smiths' home and the Greens' homes. Other small appliances are not included in our model for simplicity[a]

Equipment	The Smiths	The Greens	Average usage pattern
Lights	Six old halogen kitchen lights. Each light is 50 W	Six energy-efficient LED kitchen lights. Each light is 5 W	Four hours per day
	Three old living room lights. Each light is 50 W	Three energy-efficient living room lights. Each light is 5 W	Four hours per day
	Five old bathroom lights. Each light is 50 W	Five energy-efficient bathroom lights. Each light is 5 W	One hour per day
	Six old bedroom lights. Each light is 50 W	Six energy-efficient bedroom lights. Each light is 5 W	Four hours per day
	One old 50-W hall light.	One 5-W energy-efficient hall light	Four hours per day
	Three old landing lights. Each light is 50 W	Three energy-efficient landing lights. Each light is 5 W	Four hours per day
Clothes dryer	Clothes dryer (old) using 393 units of electricity a year	Clothes dryer (energy efficient) using 208 units of electricity a year	Regular use
Dishwasher	Dishwasher (old) using 413 units of electricity a year	Dishwasher (energy efficient) using 257 units of electricity a year	Regular use
Refrigerator	Refrigerator (old) using 400 units of electricity a year	Refrigerator (energy efficient) using 112 units of electricity a year	Regular use
Freezer	Freezer (old) using 543 units of electricity a year	Freezer (energy efficient) using 152 units of electricity a year	Regular use

(Continued)

Equipment	The Smiths	The Greens	Average usage pattern
Washing machine	Washing machine (old) using 335 units of electricity a year	Washing machine (energy efficient) using 177 units of electricity a year	Regular use
Television	Television (old) using 125 units of electricity a year	Television (energy efficient) using 25 units of electricity a year	Regular use

a Energy consumption is taken from online retailers and I try to compare like for like appliances as much as possible.

Reference

Schlömer, S., Bruckner, T., Fulton, L., Hertwich, E., McKinnon, A., Perczyk, D., Roy, J., Schaeffer, R., Sims, R., Smith, R., Wiser, R., 2014. *Annex III: Technology-specific Cost and Performance Parameters. In: Climate Change 2014: Mitigation of Climate Change. Contribution of Working Group III to the Fifth Assessment Report of the Intergovernmental Panel on Climate Change,* Cambridge, UK and New York: Cambridge University Press.

INDEX

Printed in the United States
by Baker & Taylor Publisher Services